PHILOSOPHICAL EXPLORATIONS

A Series Edited by
George Kimball Plochmann

ARITHMETIC
AND
COMBINATORICS

Kant and His Contemporaries

by GOTTFRIED MARTIN

With an Appendix
Examination of Kant's
CRITIQUE OF PURE REASON
Part I, Section 4
by Johann Schultz

Translated and Edited by Judy Wubnig

Foreword by George Kimball Plochmann

Southern Illinois University Press

CARBONDALE AND EDWARDSVILLE

Originally published as
Arithmetik und Kombinatorik bei Kant
Copyright © 1972 by Walter de Gruyter & Co., Berlin

Library of Congress Cataloging in Publication Data
Martin, Gottfried, 1901–
Arithmetic and combinatorics.

(Philosophical explorations)
Translation of: Arithmetik und Kombinatorik bei Kant.
Bibliography: p.
Includes index.
1. Arithmetic—History.   2. Combinatorial analysis—
History.   3. Kant, Immanuel, 1724–1804.   I. Wubnig,
Judy.   II. Title.   III. Series.
QA109.M3713   1985   513′.09   84–5476
ISBN 0–8093–1184–4

Contents

Foreword

Iɴ ᴀ ᴍᴏᴛɪᴏɴ ᴘɪᴄᴛᴜʀᴇ directed by Alfred Hitchcock many decades ago, the heroine, having wandered through an inconclusive love affair, now finds her affections thoroughly engaged by a much more prepossessing young man. As she clutches her new lover for the first time, a long corridor barred by a series of doors is shown in montage, and one after another they open, letting in more and more light to flood the hallway. In Gottfried Martin's book something of this sort would happen to the lover of the history of mathematical philosophy; it is not a single illumination, but light arrives in a succession of particular insights supplied for the understanding of Immanuel Kant's peculiar place in the development of modern logic and its relation to arithmetical and geometrical principles.

Ever since Aristotle made the claim for mathematics that it was a science essentially easier than others because its elements were the simplest, men and women have pursued it with the idea that here, at least, the mind could rest in a certainty found nowhere else in the range of human knowledge. On the other hand, just what reasons could be found for this claim to certainty were open to question, and there have been endless disagreements over the foundations of the science. Because of this, parties to the argument have turned to logic or other disciplines, not being content with the seeming arbitrariness of the hypotheses as presented within mathematics alone. This has, then, carried philosophers out of its strict purviews, but they could find comfort in Aristotle's own practice of seeking for the principles and defending them, not in a special mathematical treatise but in several books of the *Metaphysics,* as well as in important chapters of the *Posterior Analytics.* In these he does not set up of a list of definitions and axioms, in the style of his quasi follower Euclid; instead one finds continuous prose (in luxuriant thickets, be it said), arguing for or against proposals for considering such-and-such to be the true elements, causes, and principles entering into mathematical science.

Much the same can be said for the work of Kant, who in his immensely energetic intellectual career never wrote what could be called an orthodox mathematical text, but who, even so, never let quantities, both numbers and spatial magnitudes, get far away from the center of his thinking.

Martin's book is a box of treasures, but it must be read with care to understand its real intent. It is a book of historical evidence, together with suggestions for interpretation of sources, not a book of concerted demonstration. In this box one finds necklaces, dubloons, rings, all in this same chest of valuables. Throughout, there is a driving interest to make clear the historical milieu in which Kant's mathematical contributions developed. The author uses Leibniz, Wolff, and others whose work was finished or at least planned before Kant was born. He also makes use of Lambert, Mendelssohn, and others roughly contemporary with whom Kant kept in communication either by letter or in the press. Of Schultz and the rest of the admiring followers of Kant, Martin takes full cognizance. When a point seems to need further clarifying in more modern terms, he is ready to refer to Gauss and Grassmann, and even Frege, Russell, and Hilbert.

To change my figure, the book is a pageant, with the small-statured but intellectually gigantic figure of Kant as the chief attraction. The *Critique of Pure Reason* is examined from many points of view, with apt citations from several other Kantian works as well, some so minor that today they are almost unread except by specialists: letters, jottings in the *Reflexionen,* and occasional table talk. On each topic in arithmetic and theory of combinations Martin asks whether Kant held one consistent view throughout his career—whether seeming discrepancies between his pronouncements were no more than apparent or were rooted in a method fundamentally defective.

Many similarities between controversies of earlier centuries and the discussions agitating our own are noted. By no means were all earlier thinkers of the same mind as Kant on mathematics, whose principal contention, of course, was that some synthetic propositions could be apodictic rather than merely contingent. Indeed, many and varied combinations of the concepts of necessity, propositional structure, and content, whether mathematical or physical, were entertained and vigorously defended by writers of that earlier time, some of them even anticipating the logical positivists, who have in breaking away tried to assure us that all analytic sentences are true not because of the nature of the facts they pretend to describe but because of the structure of the language. The Wittgensteinian language games too had their predecessors in the eighteenth century. And whether one should set up a fixed number of elements and a list of rules for their manipulation is a question that Kant debated both in his own mind and with other people; it finds its counterpart in the much more recent interest in the propriety of closed systems, a matter usually associated with the name of Kurt Gödel. Martin has concluded, on the basis of much evidence, that Kant was the thinker (whether mathematician or philosopher makes no difference) preeminently responsible for the nineteenth century movement that came to be known as the axiomatizing of mathematics. A subsidiary determination is that Kant discovered the associative and commutative axioms of arithmetic first published by Johann Schultz. It is not that there is nothing new under the sun; Kant would doubtless have considerable trouble in understanding, certainly in agreeing with, the innovative formulations of Whitehead and Russell or any of their successors. But the old problems nag at philosophers, and the two best reasons for minutely studying the history of the discipline are first to derive a stimulus from problems of the past and second to avoid any mistaken solutions that may have been made.

This book is both for the specialist in the history of science and for the educated reader. It is written by a distinguished scholar with a great admiration and sympathy for Kant (he says that the *Critique of Pure Reason* is the most cleanly worked out—*sauber gearbeitet*—of all books of philosophy, a remark which should give us a measure of his dedication). For Martin, Kant is the consummate professional, regardless of the truth of the doctrines he proposes. Yet for all that, our author has written for those without any special training. A prior mastery of the *Critique* is not essential. There are a few mathematical equations but none containing any technical difficulty. Kant's expositions of the elements were not of advanced operations but of principles, for him the true metaphysical and critical task. Memorizing the symbols developed severally by the great contributors to symbolic logic is not an obligation laid upon the reader here. Indeed, a recognition of the correctness of one of the most famous sums ever mentioned in philosophic history, that seven plus five equals twelve, would probably suffice.

The following remarks are for those not fed to satiety on prefaces apologizing for the differences between literal and literary translation. Dr. Wubnig has approached the original author, and also the many philosophers and historians whom he quotes, with an abiding respect for what they have written and the manner of their writing. On the other hand, the conventions and indulgences of German philosophical writing are usually such that an extremely faithful translation sounds like—German philosophical writing. In consequence, she has docked repetitious phrases and qualifiers one takes for granted, and in general has tried to make the translation flow in a way with which English and American readers are better acquainted, at the same time preserving as far as possible the flavor of the men quoted, most of them from so long ago; nothing longer than a phrase has been tampered with. Dr. Wubnig has also added countless bibliographical and biographical details to ease the following up of Martin's allusions and suggestions.

She has wisely appended a selection from Johann Schultz, one of the first to espouse many of Kant's chief doctrines about the character of mathematical entities and operations. Schultz, with quite a different order of presentation, has taken the highly tortuous lines of the first *Critique* and turned their teachings into cogent, straightforward German. Even while starting from the standpoint of the mathematical elements themselves rather than from their status in the mind, he comes round to most of the important Kantian conclusions. It is not that Schultz denies the so-called Copernican Revolution in metaphysics, but he cares less than did Kant which celestial body moved around which. He does introduce the familiar terms *Anschauung, empirisch, a priori, rein, Verstand,* and a few others, but only after his chief premises about quantities are laid out, and much of the later part of his exposition could survive without them. He does not anticipate Frege's complaint about the "thick fat" of psychologism in logic (or mathematics), to be sure, but it is still interesting to see how a professional mathematician would read the *Critique*. His remarks, however, about time and the infinitely extensible lines would, I think, run into collision with the Transcendental Dialectic; certainly Schultz, though generally faithful to the masterpiece of 1781, is no slavish follower.

These two translations are both extremely valuable in themselves and also as supplementary to each other, and Dr. Wubnig's painstaking rendering and annotations have, I think, made clear their importance.

GEORGE KIMBALL PLOCHMANN

Translator's Preface

T HIS IS A TRANSLATION of Gottfried Martin's *Arithmetik und Kombinatorik bei Kant,* a 1972 edition of the unrevised 1938 edition, with an added chapter VI, "Synthetic Judgment in Arithmetic."

The 1938 edition included a *Lebenslauf* (brief autobiography) of Gottfried Martin. The following more up-to-date biography by his wife, Edda Martin, is included here instead.

Gottfried Martin was born on 19 June 1901 in Gera in Thuringia, the son of the minister Wilhelm Martin. He grew up in Heringen in Hessen and attended the public elementary school there; from his thirteenth year he attended the famous Friedrich Gymnasium in Kassel. He passed the matriculation exam in 1919. In 1921, after working two years in a chemical factory in Bet tenhausen near Kassel, he started his studies in chemistry at the University of Marburg. Under the strong influence of Paul Natorp, he soon changed to the study of philosophy. He completed his studies in chemistry, physics, mathematics, and philosophy with Professors Karl Friedrich von Auwers, Clemens Schäfer, Kurt W. S. Hensel, Ernst Zermelo, and Martin Heidegger.

He was drafted into military service in 1939 and discharged in 1943. Nevertheless, he did his habilitation during the war in Cologne with Professor Heinz Heimsoeth. Because of the heavy bombing of Cologne, he had to leave the city and began lecturing in Jena in Thuringia. After the division of Germany, he and his family fled from the Russian occupation to West Germany and back to Cologne. He received offers to go to Tübingen, Munich, Hamburg and Bonn, and went to Bonn as the successor to Professor Erich Rothacker. He held the chair there as professor ordinarius from 1958 until his retirement in 1969. He died 20 October 1972.[1]

Until his death Martin was the editor of *Kant-Studien,* which he had started up again with Paul Menzer in 1953 after the lapse between 1945 to 1952; of the general Kant index from 1964; of the Leibniz index from 1968; and coeditor of

Studia Leibnitiana from 1969. He wrote numerous articles and many books, a list of which appears in Part A of the Bibliography.

I would like to thank Sally Haag, Donald MacKenzie, and Joseph Novak of the University of Waterloo, and Valerie Warrior for their help with translations of Greek and Latin; David John, Manfred Kuxdorf, Edith Krause, Hildegard Marsden and Ira Tschimmel of the University of Waterloo for help with translation of the German; Paulette Trout for help with translations of French; Ross Honsberger and James Van Evra of the University of Waterloo and Roland Häggkvist of the University of Stockholm for information about the history of mathematics; E. J. Ashworth and Don Roberts of the University of Waterloo for information about the history of logic; Richard Holmes for information about Husserl; and David Binkley of the Reference Service of the Arts Library at the University of Waterloo for his help with citations. I would also like to thank Frau Gottfried Martin for information about Gottfried Martin and the writing of this work. Above all, I am grateful to Sylvia Wubnig for her help with the English, to George Kimball Plochmann of Southern Illinois University for his editorial assistance and introductory comments, and to Hans Schneeweiss of the University of Munich for all his help with both the German and the mathematics; he went through the entire manuscript and without his help this translation would not have been possible.

In addition to the considerable help from colleagues, I have found certain translations of Descartes, Hegel, Husserl, Kant, and Leibniz particularly helpful; these I list in Part B of the bibliography. Finally, for their help in preparing the manuscript, I am indebted to Arts Computing Office at the University of Waterloo, in particular, Grace Logan and Victor Neglia. I also wish to thank Durrett Wagner of The Bookworks, Inc., for his fine copyediting and the staff of Southern Illinois University Press for all their patience and help, in particular, Dan Gunter.

A last word: I have corrected inaccuracies in some of Martin's citations and have added some which are missing from his text, though I have not been able to find all of them. I have also cited the original texts of Locke's *Essay Concerning Human Understanding* and Leibniz's *Nouveaux Essais sur L'Entendement Humain*. I have used English titles for all the works of Kant except for the *Streitschrift gegen Eberhard* and for well-known works by Descartes and Leibniz. I have put the citations as endnotes, instead of in the text as Martin does, and accordingly have added some phrases in the text to indicate the author being quoted. I have also added full names of authors and their dates when possible. Except for these changes, I have indicated all my additions to Martin's text by brackets.

Part C of the bibliography includes all the works that I myself have used as well as those cited by Martin.

A Note On Translation

I follow Gabriele Rabel and Stephan Körner in translating the German *Anschauung* by the English *perception*. *Intuition,* by which it is so frequently translated, unfortunately suggests having knowledge or having hunches, which is not what Kant means (although Kant would allow the possibility that God knows everything by intuition, or, rather, that we have the [Kantian] Idea that God has knowledge this way and not by inference). See the *Critique of Pure Reason,* B71–2. He intends something closer to the original classical Latin *intuitus, looking at,* what occurs in looking at something, sensing, which is not by itself knowledge. The discussion about language in the *Critique of Pure Reason* at A320 = B376 explains Kant's usage: "Eine *Perception,* die sie lediglich auf das Subjekt als die Modifikation seines Zustandes bezieht, ist *Empfindung (sensatio)*; eine objektive Perception ist *Erkenntnis (cognitio).* Diese ist entweder *Anschauung* oder *Begriff (intuitus vel conceptus).*"

Interpreting Kant (Gram, 1982) contains several essays on the meaning of various key Kantian terms and their translations into English. Rolf George, in his *"Vorstellung* und *Erkenntnis* in Kant" (1982) argues that *Erkenntnis* should not always be translated as *knowledge,* since it did not always have that meaning in the eighteenth century. George argues that for Kant it is often more accurately translated *reference.* When Kant says (translating the above) "an objective perception is *Erkenntnis*" and "This is either . . . *(intuitus vel conceptus),*" he does not mean that merely having either kind of perception is *knowing* anything. Rather, in both cases they refer to an object; which is precisely what Kant says in the following sentence:

> Jene bezieht sich unmittelbar auf den Gegenstand und ist einzeln, dieser mittelbar vermittelst eines Merkmals, was mehreren gemein sein kann. [The former refers directly to the object and is single, the latter refers to it indirectly by means of a feature which several things can have in common.]

Of course, sensations are for Kant *Anschauungen,* but empirical, not pure and a priori like those of space and time. Compare, for instance, Körner ([1955] 1967, 27, 59, 80, 95, passim) and also Rabel's remarks on her translation of selections from Kant (1963, xv-xvi).

Jaako Hintikka ("Kant's Notion of Intuition," 1969) argues vigorously against this translation and for the more common one, *intuition.* However, he does not consider the misleading implications of the English use of *intuition,* and his main point is not really about translation; rather that the objects of

Anschauungen are particulars, but not necessarily sensations. Moltke Gram ("The Sense of a Kantian Intuition," 1982) and Hans H. Rudnick ("Translation and Kant's *Anschauung, Verstand* and *Vernunft*," 1982) both argue vigorously against translating *Anschauung* by *perception,* since Kant uses the word *Wahrnehmung* for *perception* (in contrast to *sensation* or *Empfindung*). This is, to be sure, a problem. But the problems with translating *Anschauung* by *intuition,* as they prefer, are far greater. In English, *intuition* is simply *not* what Kant meant. Far better for the translator to have to struggle with finding a good locution for the differences between *Anschauung* and *Wahrnehmung* than continually to fight the English language. To say that time and space are the forms of all (human) *perception* is a more accurate rendering of Kant's meaning than to say that they are the forms of all (human) *intuition. Perception* is, indeed, far closer to *looking at* (*anschauen*) than is *intuition.* Perhaps only a literary genius will be able to settle this dispute!

Again, I follow Rabel in translating *Vorstellung* by the English *idea.* Kant objected to using *idea* in its usual English sense because he thought it should be reserved for its original Platonic meaning. (CPR, A312 = B368ff.) I use *Idea* when I use it as Kant thinks it should be used. (Cf. Rabel, 1963, xv.)

I translate the German *Verstand* as *intellect* or *mind,* again following Rabel (1963, xvi). Although the German translation of the title of Locke's *Essay Concerning Human Understanding* is *Versuch über den menschlichen Verstand,* Locke does not mean by *understanding* what is opposed to sensibility, as Kant usually uses *Verstand.* In English, one does not normally use *understanding* to refer to a mental faculty.

On matters of proper English usage, for example, splitting or not splitting infinitives or the use of *only,* I have consulted Fowler (1983).

JUDY WUBNIG

Preface [1938]

W HEN ONE FIRST EXAMINES the theme of "Kant and Arithmetic," one is struck by the fact that numerous students of Kant wrote textbooks of mathematics. Johann Schultz (1739–1805), a particularly productive member of this close Königsberg circle, combined the position of practical theologian with that of professor of mathematics in a remarkable way.[1]

All these mathematical textbooks from Kant's circle of students are organized in the same way: They base mathematics upon an axiomatic foundation and they place strict demands on the form of proofs. Both these phenomena, axiomatization and strictness of proof, were innovations peculiar to these textbooks; furthermore, both these fundamentals go back to Kant himself. Only by putting together many details is it possible to prove this.

I have considered ways of extending the results of the present work. Further research can take two directions. First, Kant gave courses in mathematics from 1755 to 1762, and it would be desirable to have notes of these lectures. Johann Gottfried von Herder (1744–1803) was a student in one of these courses, and I have in the meantime discovered some of his notes on them. Moreover, Kant must have continued to be concerned with mathematical problems between 1780 and 1790.[2] A great many Reflexions[3] must have arisen from this work. In any case, these Reflexions still existed for some years after Kant's death, and it would be desirable to find them. However, even if we cannot do this, Herder's notes by themselves provide sufficient confirmation of what is presented in the following work.[4]

GOTTFRIED MARTIN

Introduction

THE QUESTION about the foundations of mathematics is caught in a hopeless quarrel between formalism and intuitionism, at best bogged down in an equally hopeless attempt to mediate between the two ostensibly opposed theories.

There will be no improvement until we recognize that the problem is philosophical, that is, that in questions about the foundations of mathematics purely systematic considerations must be unfruitful from the very outset and that progress can be made only by the closest combination of systematic and historical methods.

Edmund Husserl (1859–1938) holds that geometry is a field in which the theorems of geometry can be derived from axioms in a purely formal, analytic way. Oskar Becker (1889–1964) continues with this approach and goes into it more deeply in his 1923 "Beiträge zur phänomenologischen Begründung der Geometrie." This approach is almost universally accepted. This dominant view, however, contradicts the ordinary sensibility of working mathematicians. I will content myself with citing the 1816 book review Carl Friedrich Gauss (1777–1855) wrote of the commentary written in 1814 by Johann C. Schwab (1743–1821) on Book I of Euclid's *Elements*:

> Much of the work revolves around arguing against Kant that the certainty of geometry does not depend on perception but on definition and the *Principium identitas* and the *Principium contradictionis*. Kant certainly did not want to deny that these logical devices are continually being used for expressing and relating the truths of geometry to each other; but no one who is confident about the nature of geometry can fail to recognize that these principles can produce nothing in themselves and only sterile blossoms bloom when the fertile living perception of the object itself is not in control.[1]

It is often asserted with unbelievable glibness that mathematical theorems are analytic. The following was even published in a Kant festschrift:

We wish to show that, quite in Kant's own sense and contrary to outer appearances, the expansion of knowledge which is thought to occur is only formal and not material. The expansion of mathematical knowledge consists only in a different arrangement, classification, combination, division, and joining of what already exists. First of all, to illustrate this conception by a simple image from everyday life: Everyone knows that one and the same room with the same furnishings will impress the spectator altogether differently, depending on how the furnishings are arranged and distributed, so that one can indeed easily create a desired impression, a certain style, by a suitable arrangement. Mathematics is no different. An impression of something new, of progress, can be created merely by rearranging parts.[2]

Hans Beck (1876–1942) gives a survey of the situation in his 1926 *Einführung in die Axiomatik der Algebra*. The first sentence of the foreword is characteristic: "At present, the axiomatics of algebra is by no means settled."[3]

It is unsettled in two senses. First of all, it is not unequivocally certain which arithmetical propositions to establish as axioms. In any case, a certain arbitrariness in the choice of axioms appears to be a basic difficulty for every axiomatization. Second, the axiomatization of arithmetic is unsettled in that the axiomatic and the logicist[4] schools are just as opposed to each other as before. I would identify Giuseppe Peano (1858–1932) and Ernst Zermelo (1871–1958), and in a broader sense David Hilbert (1862–1943), as representatives of the axiomatic approach; Gottlob Frege (1848–1925), Louis Couturat (1868–1914), Hans Reichenbach (1891–1953), Rudolf Carnap (1891–1970), Ludwig Wittgenstein (1889–1951), and, in a broader sense Bertrand Russell (1872–1970), as representatives of the logicist approach.[5]

We can summarize the view taken for granted by almost everyone thus: Arithmetic theorems can be derived purely by means of formal logic while geometric theorems can, indeed, be derived from axioms, but then only by means of formal logic.

The present work is intended to clarify the question of how Kant saw and dealt with these problems. The basic questions in Kantian form are: Is mathematics based on axioms? Are axioms synthetic judgments? Are theorems synthetic judgments? To be sure, in all these questions, it is not even clear what Kant thought, not to speak of whether his opinions were correct.

To begin with, we encounter the view that Kant did not understand anything about mathematics. The reply is not difficult. If Kant really did not understand anything at all about mathematics when he devoted so much work and effort to it and when he gave courses in it for fifteen semesters as a privatdozent, then he should even be struck from the list of those with no more than average talents.

But it will be said that at the very least he was mathematically unproductive.

The present work tries to show both that Kant was the first to really recognize the axiomatic character of mathematics and that he proposed numerous axioms in productive mathematical work—whether alone or in collaboration with friends. But at least, it will be said, Kant's view that mathematics is based on perception [*Anschauung*] is false. However, his view is interpreted to mean that mathematics depends on appearances. The following work tries to show that Kant was perhaps the first to require rigor of proof in today's sense and that the interpretation of perception as appearance is completely misguided, understandable only by reference to the historical context.

I cite Carl Theodor Michaelis (1852–1914), Paul Mansion (1844–1919), Louis Couturat, Erich Adickes (1866–1928), and Antoon Vloemans (1898–   ) just as examples of those who attack Kant's understanding of mathematics.[6] Michaelis says in his 1884 article on Kant's concept of number:

> But Kant can no more be considered a mathematician than Christian Wolff, perhaps even less so. Wolff at least produced a series of mathematical works. He cannot be compared to Lambert, his great contemporary, the precursor of the critique of reason, however much he surpasses him in philosophy. Kant's name is not mentioned in the history of mathematics.[7]

Paul Mansion says in his 1908 article "Gauss contra Kant sur la géométrie non euclidienne": "In the *Critique of Pure Reason* itself and elsewhere Kant has shown that he only had a very poor grasp of the elements of mathematics; he did not understand anything about what was going on in the contemporary research into the first principles of geometry."[8]

What Louis Couturat maintains in his 1904 "La philosophie des mathématiques de Kant" is equally incorrect: "In fact, there is a fundamental mistake there about the nature of individual arithmetic truths, which are all demonstrable; the only primitive or indemonstrable truths of arithmetic are the general propositions or axioms, just those with which Kant did not concern himself."[9]

And Erich Adickes writes the following about the *Losen Blätter* in his 1924 *Kant als Naturforscher*:[10]

> First a word about the *Losen Blätter* with mathematical content, which have absolutely no importance for mathematics as such or for its history. They are, however, of just that much more value as psychological documents, for they allow us to come to clear conclusions about how great Kant's mathematical talent and skills were. We cannot think very highly of either of them, according to the evidence of the *Losen Blätter*.[11]

Antoon Vloemans discusses Kant's comprehension of mathematics in detail in his 1921 dissertation, *Anschauung und Verstand in der Entwicklung von*

Kants Theorie der Geometrie unter Berücksichtigung von Descartes, Leibniz, und Gauss (Perception and Intellect in the Development of Kant's Theory of Geometry in Light of Descartes, Leibniz, and Gauss). Even though Vloemans puts his arguments within a broader context of intellectual history, they are no more correct:

> In spite of his profound penetration into the nature and significance of the exact sciences, which we have been sufficiently convinced of by now, Kant is not in the same productive, active relationship to mathematics and mathematical natural science as Plato, Descartes, and Leibniz were. Kant's mathematical and scientific efforts belong to the period of his youthful development and decreased gradually as soon as he had found his independent way to his philosophical system. Philosophy withdrew from the living tradition of development and, for Kant, devoted itself from then on to its fundamental foundations. . . .
>
> [B]ut quite astonishingly, we search in vain in his work for the slightest echo of the busy research activities of his contemporaries in the sense of a critique of the axioms. . . . Kant was the philosopher who was so thorough, yet his theory remained totally unaffected by the advances in mathematics in his time; he confined himself to a dogmatic view about the axioms of geometry. This is quite understandable, given the basic tenor of his system.[12]

It is worth noting that these strong criticisms of Kant's understanding of mathematics all stem from recent decades.[13] In the eighteenth century the charge that one did not understand mathematics was fatal; if there had been any basis for it, it would have been fully used in the polemical writings against Kant. On the contrary, it was believed for as long as a century afterward that Kant understood mathematics. For example, in his 1817 *Anleitung zum Selbstfinden der reinen Mathesis*, Friedrich Schmeisser wrote:

> When it was necessary for Kant to have the basic understanding of the nature of our knowledge—which the *Critique of Pure Reason* amply shows him to have had—in order to discover the depths of the human faculty of knowledge, he displayed a comprehensive knowledge of a Newton's theories which he had studied in his reflections about the structure of the world.[14]

Here, then, Kant is credited with a thorough knowledge of the nature of mathematics and a comprehensive knowledge of mathematical theories.

According to Emil Arnoldt (1828–1905), Kant advertised and probably gave the following courses in mathematics and physics:[15]

1755/56	Mathematics	Physics
1756	Mathematics	Natural Science based on Eberhard's *Erste Gründe der Naturlehre*
1756/57	Mathematics	Physics
1757	Mathematics	Natural Science (Eberhard)
1758	Mathematics based on Wolff's *Auszug*	Natural Science (Eberhard)
1759	Mathematica varia (topics in mathematics)	Physics (Eberhard)
1759/60	Mathematics Mechanics	
1760	Mathematics	Physics
1760/61	Pure mathematics	
1761	Arithmetic, geometry, trigonometry Mechanics, hydrostatics, hydraulics, pneumatics	Theoretical physics
1761/62	Arithmetic, geometry, trigonometry	collegium physico-mathematicum
1762/63	Mathematics	Physics
1763	Mathematics	

From then until 1788 there were ten courses in theoretical physics, four based on the text of Johann Peter Eberhard (1727–1779), five on that of Johann Christian Polycarp Erxleben (1744–1777), and one on that of Wenceslaus Johann Gustav Karsten (1732–1787).

The structure of the course in physics is clear. Kant gave a two-semester course: physics in the first semester (1761/62), entitled collegium physico-mathematicum, theoretical physics in the second semester, based on Eberhard. After 1761, the course called *physics* was given up and only one in theoretical physics was still being given rather regularly from 1765/66 to 1787/88, ten times altogether, based on Eberhard, Erxleben, and Karsten. The content of these courses is indicated by the textbooks used. They contain discussions of the essentials of the atomic structure of matter, the states of aggregates and their changes, particularly those caused by heat. It is more difficult to ascertain the content of the other course in physics. It cannot have been experimental physics in today's sense; in fact, mechanics, acoustics, and optics were considered courses in mathematics. According to the title of 1761/62—collegium physico-mathematicum—we could perhaps imagine a course based on Newton's *Principia* and Kant's own early work on the forces of masses.

Four cycles can be distinguished in the courses in mathematics.

1. Summer Semester 1756–Winter Semester 1756/57. Announcement, Summer Semester 1756: "The old course in mathematics will be continued and new ones will be started."

2. Summer Semester 1758–Summer Semester 1759. This cycle begins with the *Auszug* of Christian Wolff (1679–1754) and ends with topics in mathematics.

3. Winter Semester 1759/60 Announcement: "Pure mathematics, which I start at one hour and the science of mechanics at another, both based on Wolff." Kant may have completed these courses in this semester since both courses were set as four-hour ones.

4. Summer Semester 1760–Winter Semester 1760/61.

5. Summer Semester 1761. Again, as in the winter semester of 1759/60, four hours of arithmetic, geometry, trigonometry, and four hours of mechanics, hydrostatics.

The elementary courses at that time lasted four semesters or even longer. Kant gave them in two or three semesters so he must have made them shorter. At the same time, he went deeply into applications, using Christian Wolff's *Anfangsgründe aller Mathematischen Wissenschaften*. Arnoldt cannot find the course that he thought Kant had given in fortification and pyrotechnics, but this is due to misinterpretation. According to the biography of Kant by Friedrich Wilhelm Schubert (1799–1869), "He began a series of academic courses about mathematics and physics in the winter semester of 1755, the former based on Wolff, the latter based on Eberhard's theory of nature, and he even discussed theories of fortification and pyrotechnics with sympathetic interest."[16]

Schubert certainly does not want to say that Kant gave any special courses in these subjects but only that he also spoke with interest about this tangential part of the course. It was still known to Schubert or his unnamed informants that there were sections of Wolff's textbook on mathematics which were about fortifications and pyrotechnics.

In addition to these university courses, there were also frequent private courses in mathematics for the officers of Königsberg. If, therefore, Kant gave a four-hour course in mathematics for fifteen semesters, he must be credited with having had some knowledge of mathematics. His library shows how painstakingly he prepared these courses.[17]

Arthur Warda managed to find the catalogue of Kant's books.[18] There are twenty-seven works on pure mathematics among Kant's books. We cannot be precise about one of these since it is only identified as *Mathematical Figures*. The other twenty-six books can be classified into three groups:

1. Six works which appeared after the *Critique* and are probably gifts from the authors:

Johann Schultz, *Entdeckte Theorie der Parallelen* (Theory of the Parallels Discovered), 1784;

Schultz, *Versuch einer genauen Theorie des Unendlichen* (Essay on an Exact Theory of the Infinite), 1788;

Schultz, *Anfangsgründe der reinen Mathesis* (First Principles of Mathematics), 1790;

Johann Andreas Christian Michelsen, *Eulers Differentialrechnungen* (Euler's Differential Calculus), 1790;

Michelsen, *Geometrie in Briefen* (Letters on Geometry), 1790;

Andreas Schönberger, *Grundlinien zu einer Grössenwissenschaft in ihrer Natur dargestellt* (Outlines of a Science of Mathematics Presented in Its True Nature), 1801.

2. Twelve additional works which all appeared before 1740. Kant could have acquired these when he was a student or a private tutor:

René Descartes, *Geometria* (Geometry), ed. Schooten, 1649;

Jacob or Jacques Bernoulli, *Ars conjectandi* (The Art of Estimating Probabilities), 1713;

Willem Jacob van 's Gravesande, *Matheseos universalis elementa* (Elements of General Mathematics), 1727;

Christian Wolff, *Elementa matheseos* (Elements of Mathematics), 1713–15;

Christian August Hausen, *Elementa matheseos* (Elements of Mathematics), 1734;

Johann Jesper, *Rechenbuch* (Arithmetic Book), 1682;

Jacob Köbel, *Von künstlichem Feldmessen* (On the Art of Surveying), 1578

Georg Sarganeck, *Die Geometrie in Tabellen* (Summary of Geometry), 1739;

Gaspar Schott, *Organum mathematicum* (Organon of Mathematics), 1668;

Michael Stiefel, *Arithmetica integra* (Arithmetic of Integers), 1544

Adrian Vlacque, *Trigonometria artificialis* (The Art of Trignometry), 1633;

Johann Friedric Weidler, *Institutiones mathematicae* (Elements of Mathematics), 1718.

3. Kant may have acquired the remaining eight works especially for his courses in mathematics. First there is the German edition of Wolff and two works explaining it.

Christian Wolff, *Anfangsgründe aller Mathematischen Wissenschaften* . . . (Elementary Principles of All Mathematical Sciences . . .), 1750;

Wolff, *Auszug aus den Anfangsgründen* (Abstract of the First Principles), 1749;

Christoph Andreas Büttner, *Erläuterung der Rechenkunst* . . . , *welche sich in Wolffs Auszug* . . . *befinden* (Commentary on the Arithmetic in Wolff's *Auszug*), 1754.

Then there are five larger textbooks on mathematics which appeared during the years of his courses in mathematics:

Daniel Gottlob Rudolph, *Anfangsgründe der Arithmetik* (First Principles of Arithmetic), 1757;

Johann Heinrich or Jean Henri Lambert, *Die freie Perspektive* (Free Perspective), 1759;

Johann Samuel Lilienthal, *Beschreibung einer leichten und geschwinden Methode, den genauen Inhalt aller krummen und geradelinigen Figuren zu erforschen* (Description of an Easy and Rapid Method of Investigating the Precise Content of All Curvilinear and Rectilinear Figures), 1759;

Abraham Gotthelf Kästner, *Anfangsgründe der angewandten Mathematik* (Elements of Applied Mathematics), 1759–61;

Wenceslaus Johann Gustav Karsten, *Mathesis theoretica elementaris* (Elementary Theoretical Mathematics), 1760.

This collection shows that Kant did not acquire any more mathematical works after he stopped giving courses in mathematics. Otherwise, there would have been at least one work which appeared after 1760.

Among Kant's books there are, in addition, many works in fields no longer considered as mathematics, although they were, indeed, dealt with by Wolff in the *Anfangsgründe*. (This was overlooked by Warda.)

It is just these works which give us a picture of how Kant prepared his courses. In Vol. I of the *Anfangsgründe* Wolff deals with arithmetic, geometry, and trigonometry. Engineering follows, still in Vol. I, and then the elements of artillery and fortification in Vol. II. There are eight works in Kant's library in these fields, six of which were probably acquired for the courses:

Leonhard Christoph Sturm, *Le véritable Vauban* (The True Vauban), 1710;

Sturm, *Architectura militaris* (Military Engineering), 1755;

Laurenz Johann Daniel Succov, *Erste Gründe der bürgerlichen Baukunst* (First Principles of Civil Engineering), 1751;

Anweisung zur Kriegsbaukunst (Instructions for Military Engineering), 1757;

Carl August Struensee, *Anfangsgründe der Artillerie* (Elements of Artillery), 1760;

Frederick II, King of Prussia, *Unterricht von der Kriegskunst an seine Generals* (Instruction in the Art of War to His Generals), 1761;

Joseph der Ältere Furtenbach, *Mannhafter Kunst-Spiegel, oder Continuatio, und fortsetzung allerhand Mathematisch- und Mechanischhochnutzlich- so wol auch sehr erfröhlichen delectationen, und respective im Werck selbsten experimentirten freyen Künsten,* (Mirror of the Manly Arts, or a continuation of divers mathematical and mechanical delights which are very useful and enjoyable and with liberal arts treated in the work itself), 1663;[19]

Bernard Forest de Belidor, *Architectura hydraulica* (Hydraulic Engineering), 1740.

From these acquisitions of Kant's we can see what value he put on providing realistic material for illustrating the part of his course which dealt with military and civil engineering. Wolff deals with mechanics, hydrostatics, pneumatics, and hydraulics in the second part of his *Anfangsgründe.* Here we cannot be certain whether the individual works were intended for the mathematics or the physics course.

[Royal Society of London, Académie des Sciences, Paris, and others],[20] *Neue Anmerkungen über alle Teile der Naturlehre, aus denen englischen Transactionen, denen Gedenkschriften der Akademie der Wissenschaften in Paris, und andern mehr zusammengezogen und gesamelt. Zweiter Theil. Aus dem französischen übersetzt.* (New Observations on All Parts of Natural Philosophy), 1754;

Edmund Dickinson, *Physica vetus* (Classical Physics), 1703;

Johann Christian Polycarp Erxleben, *Anfangsgründe der Naturgeschichte* (Elements of Natural History), 1768;

Stephen Hales, *Statik der Gewächse . . . mit einer Vorrede . . . von Wolff* (from the English *Vegetable Staticks,* with a preface by Wolff), 1748;

Michael Christoph Hanov, *Philosophiae naturalis sive physicae dogmaticae* (Of Natural Philosophy or Doctrines of Physics), 1762;

Jean Jacques Dourtous de Mairan, *Abhandlung vom Eise* (Treatise on Ice, from the French *Dissertation sur la Glace*);

Pierre Louis Moreau de Maupertuis, *Versuch von der Bildung der Körper*

(Essay on the Formation of Organized Bodies, from the French *Essai sur la formation des corps organisés*), 1761;

Johann Gottschalk Wallerius, *Mineralogie* (Mineralogy), 1763;

Johann Peter Eberhard, *Beiträge zur Mathesis applicata* (Discourse on Applied Mathematics), 1757 (Commentary on Wolff's mechanics and optics);

Johann Gabriel Doppelmayr, *Mathematische Instrumente* (Mathematical Instruments), 1717;

Leonhard Euler, *Mechanica* (Mechanics), 1736;

Johann Friedric Weidler, *Tractatus de machinis hydraulicis* (Treatise on Hydraulic Machines), 1728;

Andreas Werckmeister, *Orgelprobe* (Experiments with the Organ), 1716.

The third volume of the *Anfangsgründe* deals with optics, catoptrics, dioptrics, perspective and spherical trigonometry.

Daniel Bernoulli, *Hydrodynamica* (Hydrodynamics), 1738;

Isaac Newton, *Optice* (Optics), 1719;

Newton, *Philosophiae naturalis principia mathematica* (Mathematical Principles of Natural Philosophy), 1714;

Carol Benvenuti, *Dissertatio physica de lumine* (Dissertation on the Physics of Light), 1761;

The following works are about astronomy:

Pierre Gassendi, *Institutio astronomica* (Elements of Astronomy), 1647;

Johann Hevel, *Prodromus astronomiae* (Introduction to Astronomy), 1690;

Johann Leonhard Rost, *Astronomisches Handbuch* (Handbook of Astronomy), 1726;

Conrad Theophil Marquardt, *Elementa astrognosia* (Elements of Astronomy), 1734;

Jean le Rond d'Alembert, *Reflexions sur la cause générale des vents* (Reflections on the General Causes of the Winds), 1747;

Christian Gottlieb Kratzenstein, *Von dem Einfluss des Mondes* (On the Influence of the Moon), 1747;

Armand Henri Baudouin de Guemaduc, *Entdeckung eines Trabanten der Venus* (The Discovery of a Satellite of Venus, from the French *Remarques sur une quatrième observation du satellite de Venus)*, 1761;

Johann Heinrich or Jean Henri Lambert, *Cosmologische Briefe* (Letters on Cosmology), 1761;

Pierre Louis Moreau de Maupertuis, et al., *Meridiangrad, woraus man die Figur der Erde herleitet* (Degree of the Meridian from Which the Shape of the

Earth is Derived, from the French *La Figure de la Terre determinée par les observations de Messieurs De Maupertuis, . . .),* 1742;

Meteorologia (Meteorology), 1744;

Christian Gottlieb Semler, *Astrognosia Nova* (The New Astronomy), 1742;

Galileo Galilei, *Systema cosmicum* (Dialogue Concerning Two World Systems), 1699.

The conclusion of the third volume of the *Anfangsgründe* deals with geography, chronology, and gnomics. The following are about chronology and gnomics:

Johann Ulrich Müller, *Beschreibung aller derzeit üblichen Sonnenuhren* (A Description of All Contemporary Sundials), 1712;

Henry Sully, *Von der Einteilung der Zeit* (On the Division of Time, from the French *Règle artificielle du temps. Traité de la division naturelle et artificielle du temps, des horloges et des montres de differentes constructions, de la manière de les connoître et de les regler avec justesse*), 1754.

This inventory of works which Kant acquired for his library shows that he followed Wolff's *Anfangsgründe* in his course, taking special care with the parts which are no longer considered as mathematics. In particular, we see that Kant prepared richly illustrative material for the course in mathematics. We see from the list of courses and from the inventory of the books he purchased that Kant devoted a considerable amount of his time and attention to mathematics as it was commonly understood at the time.

PART I

Chapter One

The Axiomatics and Logic of Mathematics

A SHORT TECHNICAL DISCUSSION seems to be in order because the problem is difficult. I think it is just impossible to maintain so blandly and blithely that modern developments show Kant was either right or wrong—depending on the view of the author—merely by citing any passage at random from the work of someone in the field, such as Henri Poincaré (1854–1912) or Bertrand Russell. It must be pointed out over and over again that there is no agreement about the axiomatics and logic of mathematics. Moreover, the works of the creative investigators contain so many reservations that they must not be used without qualifications. Although Russell is continually cited on the possibility of a purely logical derivation of mathematics, he himself is much more cautious. He does, indeed, claim to have derived mathematics from logic, but expressly leaves open the possibility that the problems have thereby only been shunted off into logic. Moreover, it is scarcely possible to be certain about the meaning of his axiom of reducibility. Although Luitzen Egbertus Jan Brouwer (1881–1966) does establish a kind of intuitionism, the relation to Kantian *Anschauung* remains an open question.[1] David Hilbert does, indeed, begin his *Grundlagen der Geometrie* of 1899 with a citation from Kant.[2] Nevertheless, it is still questionable whether his formalistic foundation really conforms to Kant's. Although in his 1902 *Science et l'Hypothese* Henri Poincaré interprets the law of complete induction (i.e., the inference from n to $n + 1$) as synthetic judgment,[3] his interpretation is not only controversial but we may even ask whether Poincaré has used *synthesis* as Kant does. Given this situation, it seems necessary to me, first of all, to present the most important axiom systems and then to discuss the most important of the possible interpretations. Thus we can at least evaluate the consequences of any particular point of view, although we cannot really expect agreement about these questions.

The classical starting point for all axiomatics is the *Elements* of Euclid (fl. 300 B. C.). The edition of Johan Ludvig Heiberg (1854–1928) begins with thirteen propositions altogether. The editions of Euclid from the Middle Ages almost always give these tables of axioms in expanded form. Thus the 1591 edition of Christopher Clavius (1537–1612), which was probably the one used most, begins with twenty-four propositions instead of thirteen. For our purposes, we will consider Wolff's table of axioms in the *Elementa matheseos universae of* 1713–1715, where he states the following four propositions:

Axiom I: A thing is equal to itself.
Axiom II: Homogeneous quantities are either equal or unequal.
Postulate I: It is possible to draw a straight line from any given point *A* to any given point *B*.
Postulate II: It is possible to extend a straight line segment *AB* in both directions.[4]

The most important modern tables of axioms to consider are Giuseppe Peano's table in his *Arithmetices Principia* of 1889, the tables of axioms of David Hilbert, and the table of axioms in *Principia Mathematica* of 1910–13. I take Kant's list of axioms from the 1790 *Anfangsgründe der reinen Mathesis* by Johann Schultz. If the axioms in it are put together, the following table results:

Axioms
GENERAL ARITHMETIC

1. The quantity of the sum is always the same, whether the second quantity be added to the first or the first quantity to the second, i.e., it is always the case that $a + b = b + a$, e.g., $5 + 3 = 3 + 5$.

2. The quantity of the sum is always the same, whether to the first given quantity another be added either as a whole or by each of its parts, one by one, i.e., it is always the case that $c + (a + b) = (c + a) + b = c + a + b$.

GEOMETRY

1. There is only one infinite space, i.e., extended without limit on all possible sides.

2. Space is continuous, i.e., all parts of it are so related that the boundary or the termination of one part is always at the same time the boundary or beginning of the other.

3. The order in which the parts of space are next to each other is determined unalterably.

4. Geometrical surfaces and lines are continuously extended quantities, just as the volumes of physical space and geometrical bodies are. Points, however, are not extended and are the final limit of all extended things in space.

5. From a point *A* to another, *B*, no more than one straight line *AB* is possible.

6. It is not possible to extend more than one straight line through two points *A* and *B* without limit on both sides.

7. No more than one plane can go through a straight line *AB* and a point *C* not on this line without limit.

8. It is not possible for one part of a straight line to lie in one plane and the other part of it to lie outside that plane.

Postulates

GENERAL ARITHMETIC

Postulate 1: By putting several similar given quantities together successively they can be converted them into one quantity, i.e., into one whole.

Postulate 2: It is possible to augment and diminish every quantity in thought without limit.

GEOMETRY

Postulate 1: It is always possible to draw a straight line from a point *A* to another point *B*.

Postulate 2: Every straight line segment *AB* can be extended without limit through its end points on both sides *C*, *D*.

Postulate 3: It is always possible to draw a plane through a straight line *AB* and a point *C* not on it.

Postulate 4: Every plane can be extended without limit in all directions along any straight line which lies in it.

Postulate 5: In every plane it is always possible to describe a circle *ABDA* around the center *C* with the given radius *CB*.

Postulate 6: It is possible to describe a cylinder on the given base *AEBA* with the parallelogram *FGCA* perpendicular to it.

Postulate 7: It is possible to describe a cone on the base *AEBA* with the triangle *CDA* perpendicular to it.

Postulate 8: It is possible to describe a sphere around the center *C* with the radius *CA* or with the diameter *AB*.

In this present work I present new material relevant to the problem by relying on a fact hitherto unnoticed, namely, that a whole series of mathematical works, particularly textbooks, were written by Kant's circle of close followers: his friends, doctoral students, and undergraduate students. A thorough investigation of these works shows that they differ fundamentally from the other textbooks of the time: they all use an axiomatic foundation of mathematics. We can see from this material that Kant was the first to recognize the axiomatic character of mathematics and can also see the role this knowledge played in his philosophy. The following work will concentrate above all on making the relation to arithmetic clear.

For these Kantian tables I have taken the axioms from Schultz's 1790

Anfangsgründe der reinen Mathesis as Kant's. The evidence for this will be presented in the following chapters, bit by bit. Two points of view can be contrasted in comparing the tables. Mathematics according to Wolff really conforms to the ideal of mathematics without axioms held by Gottfried Wilhelm von Leibniz (1646–1716). If Kant did recognize clearly that such a foundation of mathematics was inadequate, whether according to Leibniz or to Wolff, modern developments have actually shown him to have been right. An initial comparison with Peano or Hilbert is sufficient to see that mathematics cannot be based on the assumptions made by Wolff. Moreover, a comparison with Euclid will show that no true axiomatics in today's sense can be attributed to him. Without going further, this can be seen when Euclid is compared to Hilbert or Peano. It is difficult to say what kind of meaning the propositions at the beginning of *The Elements* had for Euclid—Euclid himself, in fact, does not use the word *axiom*.[5] It appears, however, that these propositions were intended to give the definition of a geometry with compass and straightedge rather than an axiomatic system in the modern sense. In any case, Euclid probably had no idea that an axiom is a proposition whose opposite is also logically possible.[6]

Purely historically, mathematics can be divided into arithmetic and geometry. Aristotle recognized the significance of this division for the ontological problems of mathematics. A second kind of distinction is found by dividing mathematical propositions into principles and theorems. I will call the view which presupposes that there are unprovable principles *axiomatic*. I will call the opposing view *logicist*. This implies that the view that geometry is axiomatic is completely compatible with the view that arithmetic is logicist. The opposite point of view is hardly conceivable since an axiomatic arithmetic certainly requires an axiomatic geometry. Then we can take the view that the theorems can be derived from these principles purely logically. I call such a view *deductive*. On the other hand, I call the view that theorems cannot be deduced from principles by pure logic *constructive*. Even here one can hardly think of any possibilities. There are sixteen purely theoretical possibilities.[7] When those which cannot be practically realized are eliminated, there are five actually possible.

1. Arithmetic and geometry depend on axioms and are constructive in structure. Kant

2. Arithmetic and geometry depend on axioms but are deductive in structure. Jakob Friedrich Fries (1773–1843), Husserl, Hilbert, Peano, Zermelo, Johann Friedrich König (1798–1865)

3. Arithmetic and geometry can be deduced purely logically, both principles and the theorems. Leibniz, Wolff, Hermann Günter Grassmann (1809–1877), Russell, [Alfred North Whitehead (1861–1947),] Wittgenstein

4. Arithmetic is logically deductive, geometry is axiomatically constructive. Practically all the great mathematicians of the nineteenth century followed Gauss in making such a distinction between arithmetic and geometry. I mention here only Bernhard Riemann (1826–1866) and Hermann Hankel (1839–1873), and in addition, Leonard Nelson (1882–1927) and Hermann Helmholtz (1821–1894)

5. Arithmetic is logically deductive, geometry axiomatically deductive. Frege, Vloemans

The detailed classification of those who work in this area remains problematic here because these two questions are seldom explicitly formulated with precision. Accordingly, the axiomatic interpretation leaves open a possibility which the logicist interpretation really has to deny. The development of set theory shows that areas always appear at the boundaries of fields not hitherto included in mathematics. For example, while the classification of all numbers into odd and even divides the field into two parts dealt with similarly, the divisions into rational and irrational numbers, continuous and discontinuous functions, consistent and inconsistent sets, have an entirely different character. These classifications delimit determinate fields which can be understood mathematically by means of specific basic characteristics—rational numbers, continuous functions, consistent sets. Such classifications first group everything left over merely with a name and thus do not determine whether or how what is left over can be understood mathematically. An axiomatic interpretation could allow such a situation for mathematics as a whole; it could thus support the view that at times only particular parts of the field can be understood mathematically, while almost nothing can yet be said about the rest of the field at all.

We can easily see that the terms we use, *axiomatic, logicist, constructive, deductive,* correspond to the Kantian terms *analytic* and *synthetic,* since the axiomatic interpretation uses synthetic principles and the logicist interpretation analytic ones. The constructivist interpretation also requires synthesis for the derivation of theorems in contrast to the deductive interpretation, according to which theorems are derived by analytic judgments. In order to make these divisions clearer, I will refer to what some in the field have said.

Leibniz and Wolff argue quite unequivocally for a purely logical deduction of theorems as well as of principles. Leibniz gives proofs for all twenty axioms (as according to Clavius) in his observations on Euclid, Book I.[8]

> I am certain that the proof of the axioms is of great utility for either true analysis or the art of discovery. . . . On the contrary, I am of the opinion that geometers ought to be praised because they have secured science by means of these propositions, which are like pegs, and have discovered the art of advancing and deriving so much from so little; for if they had tried to

put off discovering theorems and problems until all the axioms and postulates had been proved, we might not have geometry today.[9]

Thus for Leibniz the axioms are an imperfection in the structure of geometry which must be tolerated until proofs of them have been found.

From the list of axioms in Wolff's *Elementa matheseos universae* given above, it follows that he carried out a purely logical construction of a mathematics without axioms. However, he also says expressly, "It must, of course, be noted that the number of axioms decreases as our concepts develop more adequately. In fact, if I may speak the truth, there are no true axioms except identical propositions."[10]

Johann August Eberhard (1739–1809) and Johann Christoph Schwab also took the same point of view in opposition to Kant. Eberhard says in an article in his magazine in 1789–1790, "Why can't the imagination imagine two straight lines between two given points? A higher intellect . . . would be able to find the reason for this. It would then be apodictically certain, and this certainty would be rational not sensible."[11]

Thus for Eberhard too, an axiom has an absolutely rational basis, only we do not know it. And Schwab says in his 1814 commentary on Book I of Euclid's Elements: "and in order that no geometric axiom be left among the Euclidean axioms, we shall prove axioms 10 and 12, and it will thus be clear that all geometry rests on the most fundamental principles of human knowledge."[12]

For deduction from axioms I can cite Hilbert's 1908 "Axiomatisches Denken":

When we consider any particular theory more closely we always see that a few special propositions of the area of knowledge lie at the basis of the construction of the framework of the concepts, and these are then sufficient by themselves for constructing the whole framework according to logical principles.[13]

For the ontological distinction between arithmetic and geometry, I first of all cite Gauss in an 1830 letter to Bessell:

My deepest conviction is that the theory of space occupies an entirely different position in our a priori knowledge from that of pure arithmetic; it cannot completely persuade us of its necessity (thus also of its absolute truth) as the latter can; we must humbly admit that if number is merely a product of our mind, space has a reality independent of our mind whose a priori laws we cannot completely prescribe.[14]

Frege and Vloemans make similar comments. In the 1884 *Foundations of Arithmetic*, Frege writes:

I see it as a great credit to Kant that he made the distinction between analytic and synthetic judgments. Insofar as he called the truths of geometry synthetic and a priori, he revealed their true nature. And this is still worth repeating because it is still often misunderstood. Even if Kant was mistaken about arithmetic, I do not believe that this detracts from his contribution in any essential way. What was important to him was that there were synthetic judgments a priori; whether they occur only in geometry or also in arithmetic is of minor importance.[15]

And Vloemans writes in his dissertation, "The pure sciences of intellect teach us the same thing: logic and arithmetic. . . . It is, however, different in the case of geometry where a foreign element confronts the intellect."[16]

Apart from the evidence from Kant and Schultz to be given later, Christian Gottlieb Zimmermann (1769–1841) may be cited here to show that Kant and his students adopted the axiomatics of mathematics completely aware of what they were doing and quite conscious that they were in opposition to the developments up to their time. Zimmermann writes in his 1814 *Anfangsgründe der Geometrie:*

We may completely lose all these advantages of mathematics by not handling it correctly. The method here must keep to its characteristic level and cannot deviate in rigor and its requirements. On the one hand, to want to multiply the number of principles by theorems which the intellect does not accept as being basic propositions, and on the other hand, to want to explain and prove what does not need explanation or proof and so by which the intellect gains neither clarity nor insight . . . are mistakes inconsistent with the spirit of mathematics and its methods. These remarks would be superfluous if men of reputation and authority in their writings had not often overlooked this rule, which is based on the nature of things, and by making these mistakes had not presented the whole thing inadequately. Even Proclus in his commentary on the Elements of Euclid made an effort to reduce the number of axioms. . . . Apollonius of Perga also tried something similar and, among the modern mathematicians, so did Roberval, and above all, Leibniz (*Oeuvres post.* LIV Ch. XII).[17]

Zimmermann was recognized as a favorite student of Kant's, and in his mathematical textbooks he takes exactly the same point of view as his teacher. He particularly emphasizes the very clear position he takes against Leibniz by giving exact citations for the points of disagreement.

This short survey shows that there is not the same general agreement about axiomatics as there is in other areas of mathematics. The descriptions of the axioms show essentially greater differences than do the descriptions of other areas of mathematics.

At the same time, the passages cited show that the basic problem of axiomatics is its relation to logic. Can axioms be derived by logic? Can the theorems be derived from the axioms by logic? Thus we also find the link to one of the great fundamental problems of Kant: his opposition to the then-accepted logic.

CHAPTER TWO

The Analytic Principles

IN THE COURSE of his investigations into axiomatics, Kant took some of the axioms and described them as analytic principles. The two main passages are both in the *Critique of Pure Reason*.

> [Axioms of Perception][1]
> As regards magnitude (*quantitas*), that is, as regards the answer to the question of how large something is, there are no axioms in the strict sense of the term, although there are a number of propositions which are synthetic and immediately certain (*indemonstrabilia*). The propositions that if equals be added to equals the wholes are equal, and that if equals be taken from equals, the remainders are equal, are analytic propositions; for I am immediately conscious of the identity of the production of the one magnitude with that of the other; axioms, however, have to be a priori synthetic propositions.[2]

In the *Prolegomena*, in the detailed account of synthetic judgments, Kant again considers the analytic principles. The passage was taken over into the introduction to the second edition of the *Critique of Pure Reason*, with two important changes:

> Some other (B16 few) principles presupposed by the geometer are, indeed, really analytic and depend on the law of contradiction; but they only (B16 also only) serve as identical propositions, as links in the chain of method but not as principles; e.g., $a = a, (a + b) \rangle a$, i.e., its whole is greater than its part. And, indeed, even these, although they are recognized as valid according to mere concepts, are only admitted into mathematics because they can be exhibited in perception.[3]

Johann Schultz combines the two passages in his *Prüfung der Kantischen Critik der reinen Vernunft* of 1789:

It does, indeed, at first appear as if there were no axioms of arithmetic. For all the propositions usually presented as axioms in textbooks of arithmetic are either purely analytic consequences of definitions, e.g., like the propositions that the whole is equal to its parts taken together and that no part is greater than the whole; or they are consequences of the proposition A = A, i.e., of the principle of contradiction itself, e.g., like the proposition that two quantities which are equal to a third are equal to each other; or they can actually be proved, e.g., like the propositions that equals added to equals or subtracted from them yield equals, as *Wolf* has already shown in his *Elementa matheseos universae*.[4]

Kant's explanation of these analytic principles is unambiguous. We shall make a precise survey of the propositions and their derivation. It will follow, first of all, that the claim that the passage B16ff. is perverse is completely unfounded. Later, we will meet a series of well-defined judgments which are explicitly called analytic judgments by Kant. Thus, the criticism that analytic judgment is not adequately defined by Kant and that, moreover, there are no adequate or convincing examples is equally unfounded. We will also face a difficult problem here which Kant was not, in fact, able to solve. If there is a field of analytic propositions, we can pose the question, particularly with regard to Kant, as to which science these propositions belong, mathematics or philosophy. Depending on the answer to this question, either mathematics or philosophy will acquire a discipline in no way involving synthesis but only analytic judgments.

The starting point for all discussions about axiomatics is Euclid's *Elements*. We have already indicated that Euclid himself did not use the term *axiom*. He used only *conceptiones communes* and *postulata*.[5] It is, moreover, generally admitted that Euclid did not formulate the elements himself but that he basically took them from Aristotle.[6] The significance of the axioms remains as obscure for Euclid as it does for Aristotle. Jakob Fries and Hermann Hankel have already shown that Euclid was unfamiliar with axioms in the true sense of the word. Further work stemming from Euclid's table of principles started at once to go in three different directions: (1) attempting to clarify the concepts of definition, axiom, and postulate; (2) attempting to prove the axioms; (3) attempting to complete the table of axioms by adding more of them. Proclus (ca. 410–485) reviewed these three attempts in his commentary on the first book of Euclid's *Elements*. In any case, the discussion of Euclid's table of axioms shows that neither the individual concepts of definition, axiom, or postulate, nor the propositions which are themselves considered as definitions, axioms, or postulates could be clearly made out. Christopher Clavius maintained that first of all we should seek out all the axioms;[7] on the other hand, Petrus Ramus (1515–1572) and J. A. (Giovanni Alfonso) Borelli (1608–1679) in his 1658 edition of the *Elements* denied that there are any axioms in the strict sense.[8] The two very well-known English works, the 1621 *Praelectiones tresdecim in principium Elementorum Euclidis* of Henry Savile (1549–1622), which gets as far as the seventh proposition of the first book, and the 1659 edition of Isaac Barrow (1642–1727), basically follow Clavius and so does Giovanni Girolamo Saccheri (1667–1733) in his *Euclides ab omni naevo vindicatus* of 1733.[9]

The attempt to prove the axioms was taken up again by Gilles Personne de Roberval (1602–1675), as Leibniz tells us in the *New Essays Concerning Human Understanding*: "And I remember when I was in Paris, the late M. Roberval, then old, was being laughed at because he wanted to demonstrate Euclid's axioms, following the examples of Apollonius and Proclus, and I showed the importance of this investigation."[10]

Leibniz went further with this attempt to prove the axioms, and he was completely clear about what he was doing. He not only tried to prove those in Euclid's *Elements* according to Clavius's list, but elsewhere he gave proofs and explanations of why these proofs were necessary: "for far from approving the acceptance of doubtful principles, for myself I would even like to see an attempt to demonstrate Euclid's axioms, as some of the ancients tried to do."[11]

Leibniz's program finds expression in the great textbooks of Christian Wolff. The reference already given to the axioms of his *Elementa matheseos universae* shows that it actually deals with a completely axiom-free, purely logicist mathematics in today's sense. Practically all the textbooks of the eighteenth century were under the influence of this approach of Wolff's and Leibniz's. However, the authors generally prefer to express themselves very carefully and tortuously so that when they do now and then present principles, they immediately add that those are provable or regret that the proof has not yet been found. On the other hand, I cannot tell whether this procedure also had the same almost complete success in England and France because of lack of material. In England, John Locke (1632–1704) would have been opposed to it since he discusses axioms in detail in an entire chapter of his *Essay Concerning Human Understanding*. Likewise, Savile and Barrow, and (this may well be decisive) Isaac Newton (1642–1727), took a completely axiomatic point of view. In Germany, the highly respected Johann Jakob Hentsch (1723–1764) followed Locke in his *Philosophia Mathematica complectens methodum cogitandi* of 1756.[12] *Die Geometrie in Tabellen* by Georg Sarganeck (1702–1743), which presents almost 100 axioms of geometry, is quite outside the mainstream.[13] It cannot be a coincidence that this work is to be found among Kant's books. And Georg Jonathan Holland (1742–1784) expresses himself very cautiously in his 1764 treatise on mathematics:

I do not deny that there are some universal ontological principles which are common to geometric and arithmetic quantities. They are very limited in number, however, and primarily concern continuous quantity only insofar as it is regarded as a divisible whole. Karsten[14] made an attempt to present these propositions so that one could base arithmetic and geometry on them as coordinated sciences. He himself later admitted that he now held the greater part of his general mathematics to be a purely abstract theory of numbers.[15]

A conscious axiomatics is first found included in the work of Schultz and in that of Kant's immediate students: Johann Gottfried Karl Christian Kiesewetter (1766–1819), Christian Gottlieb Zimmermann, and Jacob Sigismund Beck (1761–1840). Some attempts to come to a systematic axiomatics or to maintain a single basic axiom are directly connected to Kant. In the first place, Jakob Fries must be mentioned here; he deals with arithmetic and combinatorics almost exclusively in his 1822 *Die mathematische Naturphilosophie*, ignoring geometry. He establishes one supreme axiom in order to derive all other axioms from it.

> The supreme form of all axioms is: It is always the case that between given boundaries one and only one part of the series is possible. . . .
> (Axiom of Arithmetic). From homogeneous quantities as parts, one and only one sum is possible. Addendum 3: A quantity as a whole can always be set as equal to any arbitrary combination of all its parts, for only one quantity is possible from these parts. All complexes from the same arithmetic elements are equal to each other.[16]

Friedrich Schmeisser argues for a similar position in his 1817 *Lehrbuch der reinen Mathesis*:

> A thing *A* must either have the predicate of another *B* or not; there is no third possibility. As applied to quantities, this proposition is as follows: a quantity *A* must either have the quantity of another *B* or not; in other words, *A* must either be equal to *B* or unequal. Thus we have the proposition which satisfies all the above requirements and, as a true axiom, constitutes, as it were, the highest point of our science. . . .
> Now how arithmetic develops systematically from this proposition is shown.[17]

And Alois (Aloys) Mayr (1807–1890) also takes this view in his 1845 *Untersuchung über die wissenschaftliche Methode*.

> The next thing that we see precisely is that there can be one, and only one, axiom in any kind of fundamental science. . . .
> The mathematical axiom will rather have to be sought in the second

proposition, namely, that we get the same result whatever order the individual quantities are added in. . . .

Then the axiom will be: two quantities added together again result in a quantity . . . whatever order they may be added in. . . .

If this proposition is formulated as the true mathematical axiom. . . .[18]

It is interesting that in these attempts the alleged basic axiom is always presented as an amalgamation of the associative and commutative laws of addition. Such attempts to develop mathematics are naturally found more often after Georg Wilhelm Friedrich Hegel (1770–1831). None of them is, however, quite to the point, and they all fail to capture the basic meaning of Kantian axiomatics. By far the most important of the attempts depending on Hegel is the work of Hermann Grassmann, which we will deal with more fully in what follows.

Equality is (1) reflexive, $a = a$; (2) symmetrical, from $a = b$, $b = a$ follows; and (3) transitive, from $a = b$ and $b = c$, $a = c$ follows. We can choose two ways for dealing with it. We can define equality and then try to derive the three properties of equality from the definition, or we can simply define equality by its properties. The first method is the logicist one, the second the axiomatic. Peano chooses the purely axiomatic way. He lays down the three properties of equality in his Axioms 2, 3, and 4.[19]

Generally, however, the purely logicist way is chosen for dealing with equality where the latter is always defined in some sense as substitutability. Leibniz defines equality as substitutability; according to him, we can do without the property of reciprocity:

Definition: equals are those which can be substituted for each other without affecting truth. It seems too much to say that they are deceptive. For from the fact that the later one can be substituted for the earlier one, without affecting the truth about quantity, it follows in turn that the later one can also be substituted for the earlier one in a similar way, as I have shown elsewhere. Now to put in something by anticipating it when it can be demonstrated from definition is, strictly speaking, deceptive.[20]

Leibniz also gives practically the same proof in other places:

I formulate a theorem of this sort: if A can be substituted anywhere in place of B, B can always be substituted for A without affecting the truth. I demonstrate this with the help of an Axiom: B can be substituted anywhere in place of B. For if A can be substituted anywhere in place of B (according to our hypothesis), let it also be substituted in the later position in this Axiom: B can be substituted anywhere in place of B, hence it follows: B can be substituted anywhere in place of A. Q.E.D.[21]

Leibniz thus arrives at the following definition as his final result: "Those things are the same which can be substituted for each other without affecting the truth."[22] And Wolff uses this Leibnizian definition in his *Philosophia prima, sive Ontologia* of 1730:

> Things are called the same when they can be substituted for each other without affecting whatever was previously stated to be applicable to one of them, either absolutely or under a given condition.[23] Things which can be substituted for each other without affecting their quantity are equals. The opposite: things which cannot be substituted for each other without affecting their quantity are unequal.[24]

Leibniz and Wolff, therefore, define equality as identity of quantity, the unequal as the not equal.

Schultz, on the contrary, in the *Anfangsgründe* defines the equal the other way around, as that which has no distinctions. The difference may appear to be unimportant. In spite of this, however, it deserves to be kept in mind.

> Explanation 1: The internal characteristics of a thing are those concerning its quality and quantity, the external those which indicate its position in space and time.
>   Explanation 2: Things which cannot be distinguished by internal characteristics are, considered in themselves, completely identical.
>   Explanation 4: Homogeneous things with the same quality are called similar (*similia*), with the same quantity equal (*aequalia*).[25]

Hermann Grassmann in his 1844 *Die Wissenschaft der extensiven Grössen oder die Ausdehnungslehre* goes back to Leibniz's definition again:

> Thus we have saved the simplicity of the concept of equality and can define it as follows: Those things are equal of which we can always say the same thing, or, more generally, which can be substituted for each other in every judgment. As is implicit here, it is clear that when two forms are equal to a third, they are also equal to each other, and that what is derived in the same way from equals is again equal.[26]

Every logicist treatment of equality has then to derive the three properties of equality from the respective chosen definitions. The usual proof is in two parts. The reflexivity of equality is presented as a particular case of identity. Symmetry and transitivity are then derived from this proposition and together with the definition of equality. Leibniz presents the proposition $a = a$ as an identical proposition without further ado. Accordingly, in his *Elementa matheseos universae* Wolff holds this proposition to be identical, but as an axiom: Axiom 1: A thing is equal to itself.[27]

The derivation given by Wolff in the *Ontologia* (section 351) can be ignored for our purposes. Johann Heinrich (Jean Henri) Lambert (1728–1777) in his 1771 *Anlage zur Architektonik* also presents the proposition—every quantity is equal to itself—as an identical proposition immediately taken to be true as soon as the words are merely understood.[28] By distinguishing between arbitrary and genuine principles, Lambert here prepares for the distinction between analytic and synthetic propositions by distinguishing between analytic principles and synthetic axioms.

Schultz, following Leibniz and Wolff, traces the reflexivity of equality back to identity as a special case: "Addendum 6: Every quantity is equal to itself, i.e., $a = a$. For according to either the principle of contradiction or that of identity the quantity a is the quantity a.[29]

The symmetry of equality was, indeed, always known and always used, but, so far as I know, was first explicitly formulated in the nineteenth century. The definition just cited from Leibniz explicitly entered into the question of whether symmetry must be introduced into the definition.

On the other hand, the transitivity of equality was explicitly formulated as a principle even by Euclid, and the attempts at proof have always begun with just this principle. The attempts began with Apollonius of Perga, about two centuries after Euclid. The proof is preserved by Proclus,[30] and, together with Proclus's critique which follows it, was taken over by Savile in his lectures on the *Elements*.[31]

The idea of a proof as really amounting to a substitution was then also continued by Leibniz, Wolff, and Kant. In discussing Book I of Euclid's *Elements*, Leibniz gives the following proof: "Let a and b be equal, and let c be equal to a. Then I say that c will also be equal to b. For let $a = c$ since $a = b$, then b can be substituted for a without changing its quantity and b will be equal to c. Q.E.D."[32]

Wolff gives the same proof for equality in the *Elementa matheseos universae* and in the *Ontologia*, and completely generally for identity in the *Ontologia*.[33] Schultz also uses the same approach for proving this:

Addendum 7: (On the definition of equality). If every one of a group of several quantities is equal to one and the same quantity, they are equal to each other; i.e., if $a = x, b = x, c = x$, etc., then $a = b = c$, etc., because $b = x$ and $c = x$. Thus, in the proposition $a = x$, either b or c can replace x, and in the proposition $b = x$, c can likewise take the place of x.[34]

In spite of the agreement between Kant and Leibniz, the purely logicist treatment of equality cannot be completely persuasive. Helmholts shows in his 1887 "Zählen und Messen, erkenntnis theoretisch betrachtet," that in almost all concrete cases the definition of equality as substitutability makes no sense.[35] Even equality of length is not determined by substitutability but by laying the lengths in question one over the other. If we try to define the equality of two temperatures in terms of substitutability, the definition must really be stretched in order to get even a halfway meaningful sense. Federigo Enriques (1871–1946) comes to the same conclusion in his 1927 *Zur Geschichte der Logik*:

> If the equality of temperature is defined positively by an experiment with temperature equilibrium, it still cannot be maintained here a priori that "two temperatures equal to a third are" equal each other. To speak as Kant does, what appears to be an *analytic* judgment is really a *synthetic* one.[36]

The last proposition, of course, can only be stated with caution. There is no doubt that the propositions about equality were analytic principles for Kant, that is, that he thought they could be proved by means of definitions and the Law of Identity or of Contradiction. If we were persuaded that the proofs of these propositions were not foolproof, then we would, of course, have to omit them from the class of analytic principles and they would then become synthetic axioms in the Kantian sense.

Even Russell's attempt to prove the propositions about equality with the most powerful tools cannot be completely persuasive. The question is posed here of how far the consequences, in particular the symmetry and transitivity of equality, are implicit at the start. If they are, this would imply that the axioms are merely shunted off into logic, a possibility which Russell, who was extremely cautious, himself leaves open. Helmholtz's objections would, however, be doubly powerful because if it is already difficult to define the equality of two temperatures in terms of substitutability, it would seem to be quite impossible to subsume the equality of two temperatures under Russell's definition of equality.

Both the main passages have shown us that for Kant the propositions of equality are analytic principles. Schultz's work in the *Anfangsgründe der reinen Mathesis* makes the significance of this definition clear: Analytic principles must be provable from definitions and the Law of Identity or of Contradiction. The passages in Kant which will be cited later show that this was precisely Kant's view. If we were to be persuaded, in opposition to Kant, that the three principles of equality cannot be proved, then these propositions could not be included in the class of analytic principles and would have to be regarded as genuinely synthetic axioms.

The proposition that equals added to equals give equals comes right after the basic laws of equality. The proposition can then be modified for all other operations. Wolff gives the proof in the *Elementa matheseos universae* (section 88) and in the *Ontologia* (section 371). The Wolffian proof is in agreement with

Leibniz's here as well. According to a remark of Leibniz's, Roberval began his attempts at proving the axioms with just this axiom.[37] Leibniz gives the following proof in his remarks about the first book of Euclid:

> All this can be demonstrated from the definition of equal things. . . . See Axiom 2. Let $a = 1$ and $b = m$. Then $a + b$ will be equal to $1 + m$. For $a + b = a + b$; 1 and m are substituted on either side of the equation for a and b, and then $a + b$ will be equal to $1 + m$.[38]

Schultz completely follows this method of proof of Leibniz and Wolff:

> Addendum to explanation 7: If equals be taken together with equals then the resulting quantities are equal; i.e., if $a = b$, $c = d$, then $a + c = b + d$. For in the proposition $a + c = a + c$, b can be substituted for a and d for c . . . hence also $b + d$ for $a + c$. Therefore $a + c = b + d$.[39]

Peano presents this proposition as theorem 22.[40] Schultz's proof uses the inference from n to $n + 1$; the restriction which results can be equally valid for Peano since he only has the foundations of arithmetic in mind. For this construction, however, Peano uses the axioms he lays down, some of which are synthetic axioms in the Kantian sense. If Peano's method of proof were the only one possible, then the proposition that equals added to equals give equals would not be an analytic principle in the Kantian terminology but a synthetic theorem. For Russell the observation already made is valid. Locke goes into great detail about this very proposition in the chapter about axioms. Since this passage is characteristic of Locke's way of thinking, I will give it here:

> *Thirdly,* as to the *relations of modes,* mathematicians have framed many axioms concerning that one relation of equality. As, "equals taken from equals, the remainder will be equal"; which, with the rest of that kind, however they are received for maxims by the mathematicians, and are unquestionable truths, yet, I think that any one who considers them will not find that they have a clearer self-evidence than these, that "one and one are equal to two"; that "if you take from the five fingers of one hand two, and from the five fingers of the other hand two, the remaining numbers will be equal." These and a thousand other such propositions may be found in numbers, which, at the very first hearing, force the assent, and carry with them an equal, if not greater clearness, than those mathematical axioms.[41]

Surely Locke can be justly criticized here by using the point that Kant made repeatedly: in mathematics a proposition which can be proved must never be taken as unprovable.

Euclid ends the list of his common notions with the proposition "The whole is greater than the part." Later, others (e.g., Clavius) added the proposition closely related to it: "The whole is just as large as the sum of its parts." Wolff proves both propositions in the *Elementa matheseos universae* and the *Ontologia*. His two proofs do not basically differ; I cite the proof from the *Ontologia*. Wolff first proves the second proposition after giving definitions of *greater, smaller, whole,* and *part*.

> One thing which is identical to many is called a *whole*; conversely, many things which if taken together are identical to one thing are called *parts* of it and each one of them is called a *part*.[42] If any part of one thing is equal to another whole thing, then the former is greater than the latter; but if one thing is equal to a part of another, it is less than the latter.[43]

Since according to this definition the totality of the parts is altogether identical to the whole, we can immediately make the inference about partial identity, namely, identity of quantity:

> All the parts taken together are equal to the whole. For all the parts taken together can be substituted for the whole without changing anything previously stated which was attributed to it and consequently without changing its quantity.[44]

As for the first proposition, Wolff proves at the outset that a part of a whole is smaller than the whole. Through inversion he then derives the desired proposition easily:

> Any part of a whole is less than the whole. For any part of the whole is equal to itself and for that reason, according to our hypothesis, it is equal to a part of the whole. It is therefore less than the whole. The demonstration of the present proposition provides an example of a complete analysis since the principles which it depends on are the definition and the axiom properly so called, that is, an identical proposition.[45]

We can doubt whether this proof is really conclusive. The method of proof stems from Leibniz.

> If a part of a quantity is equal to another quantity as a whole, then the first quantity is called greater, the second lesser. Hence the whole is greater than the part.
>   Given the whole *A*, the part *B*, I maintain that *A* is greater than *B* because a part of *A* (namely *B*) is equal to the whole of *B*. This can also be expressed by means of a syllogism with the definition as the major premise and an identity as the minor, namely:
>   What is equal to a part of *A* is less than *A*, according to the definition.

Now B is equal to itself so, according to the hypothesis, it is equal to a part of A.

Therefore B is less than A.

Here we see that all proofs ultimately depend on two kinds of unprovable foundations: on the definitions or ideas and on primitive propositions or identities such as B is B, that everything is equal to itself, and innumerably many others of the same kind.[46]

Leibniz argues in the *New Essays*, against Locke's position. Hentsch, for his part, returns to Locke again in his *Philosophia mathematica* and holds that the proposition that the whole is greater than the part is unprovable.

It is on the question of this proposition that Schultz parts from Leibniz and Wolff for the first time. He gives a proof which for the first time in mathematics, so far as I can see, takes into account the restrictions under which the proposition is valid.

Theorem I. If a part b of the whole $b + a$ is either equal to the other part a or if this latter [a] is equal to a finite part of the former [b], then each of the parts is smaller than the whole and the whole is greater than any of its parts.

Proof: For if $b = a$, then $a + a = b + b = b + a$. Now since $a ⟨ a + a$, and $b ⟨ b + b$, then on the one hand, $a ⟨ b + a$, and on the other, $b ⟨ b + a$.

If, on the contrary, it is not the case that $b = a$ and that a is a finite part of b, then $a ⟨ b$, so that to begin with, $a ⟨ b + a$. Furthermore, there is such a finite set m of a, that either $b = ma$ or $b ⟩ ma$ while $b ⟨ ma + a$. But now if $b = ma$, then $b + a = ma + a$. Now since $ma ⟨ ma + a$, it is also the case that $b ⟨ ma + a$, so that $b ⟨ b + a$. On the other hand, if $b ⟩ ma$ and $b ⟨ ma + a$, then ma is a part of b; accordingly, $b = ma + p$, hence $b + a = ma + p + a$. Now since $b ⟨ ma + a$, then $b ⟨ ma + p + a$; it follows again that $b ⟨ b + a$. Therefore in all cases, not only is $a ⟨ b + a$, but also $b ⟨ b + a$.

Addendum: Therefore, in finite quantities, the part is always smaller than the whole and the whole greater than the part.

Note: The proposition that the whole is greater than the part is so obvious with respect to finite quantities that it has always been presented as an axiom which needs no proof. But since the basis for the certainty about it can still be demonstrated, I feel obligated to demonstrate it and thus to demonstrate the proposition as a theorem and, indeed, the more so, since one is used to imagining that it must not only be valid for all finite quantities but also generally for all infinite quantities.[48]

For this kind of proof, the subordinate propositions and the note show clearly that Schultz was consciously in total opposition to Leibniz and Wolff. While for Leibniz and Wolff the proposition that the whole is greater than the part is absolutely and unconditionally valid, according to Schultz it is only valid for finite sets. Or another way of putting it is that Kant and Schultz have turned the proposition that the whole is greater than the part into the proposition that the sum is greater than any of its terms. Fundamentally, these ways of thinking lead to the problems of modern set theory; Giulio Vivanti (1859–1949) showed in his 1908 "Infinitesimalrechnung" that much of modern set theory is contained in Schultz's work.[49]

This turning of the proposition that the whole is greater than the part into the proposition that the sum is greater than any of its terms explains a remarkable formulation of it which Kant gives in the *Prolegomena* and in the *Critique of Pure Reason*, B17, when he writes $(a + b) \rangle a$.

Now the whole field of analytic principles is remarkably informative for Kantian philosophy, but this last proposition is really the most interesting, both in its formulation and in its proof; and it seems very strange to me that in all the voluminous Kant literature only Louis Couturat saw the problems which arise here:

> Furthermore, one could comment that Kant chooses a rather poor example of an analytic principle, "The whole is greater than the part," which he formulates as "$(a + b) \rangle a$." In effect, this proposition is not even a principle or an axiom because it is only true of certain kinds of quantities and not of all. . . . For example, this theorem is true for the finite numbers but it no longer holds true for the infinite cardinal numbers. Doubtless one cannot criticize Kant for having ignored these truths, even if they are elementary today. But one wonders, nevertheless, how he could have admitted such a proposition as analytic, given his own principles.[50]

In examining this question it becomes clear how unjustified the criticisms are that Kant had an inadequate mathematical background. Couturat does not want to hold Kant and Schultz responsible for ignorance of the very points they had worked out!

Peano defines the concepts of *greater* and *smaller* in terms of the whole numbers. His definition amounts to the proposition that one whole number is greater than a second one when it is produced from the second by the addition of a third, or that the *later* numbers of the number series are greater than the preceding ones. This is basically the same approach that Schultz had followed.

The definition which Schultz gives of *greater* and *smaller*—"If through taking a quantity once or many times . . . "—refers back to the Euclidean definition, Book V, Definition 1. By means of this definition all the discussions in Book V dependent on it are limited to commensurable magnitudes; consistently, Euclid does not introduce the concept of incommensurable magnitudes until Book X. Schultz follows the same plan. He too first proves the proposition for commensurable magnitudes, then extends the proof to the incommensurable magnitudes with the axiom of Archimedes. Thus the basic question is clearly raised again. Is the proposition "The whole is greater than the part" derived from the ideas of *whole, part, greater,* and *smaller*, and is it thus obviously valid, or do the meanings of *whole, part, greater,* and *smaller* always have to be redefined so that the proposition also has to be proved again? Leibniz and Wolff give a proof on the basis of the ideas and so a proof having unrestricted validity. In identifying the analytic propositions as analytic principles, Kant and Schultz do indeed interpret the propositions in the same way as Leibniz and Wolff, but with the qualification made in the actual proof they depart fundamentally from Leibniz and Wolff. Couturat's criticism is thus justified.

A particularly important precritical work for our question is Kant's Prize Essay, written at the end of 1764: *Enquiry into the Clarity of the Principles of Natural Theology and Morals.* The remarks up till now have shown that the analytic principles, and only these, can be proved by means of definitions and the Laws of Identity and Contradiction. Obviously then, other sources of knowledge must still be found. Christian August Crusius (1715–1775) had already maintained that the Law of Contradiction could in no way be the only sufficient ground of all knowledge, but that, rather, material principles were necessary in addition to this formal principle. Chapters 7 and 10 of Crusius's 1747 *Weg zur Gewissheit und Zuverlässigkeit der menschlichen Erkenntnis* deal with this question: "One must . . . recognize instead that the Law of Contradiction, since it is an empty proposition, is not the only principle of human certainty."[51]

Kant refers to Crusius's view in detail and, indeed, is in fundamental agreement with it.[52] Kant leaves it open whether the material principles presented by Crusius are correct as given:

> Now in philosophy there are many unprovable propositions, as has already been mentioned above . . . but insofar as they also contain the grounds of other knowledge, they are the primary material principles of human reason. . . . Such material principles constitute, as Crusius rightly says, the foundation and the soundness of human reason.[53]

The question which is of particular interest here, whether mathematics requires material principles in addition to the Law of Contradiction and its variations as formal principles, is answered *no* by Crusius, while by 1764 Kant had answered it *yes*. Crusius says, "Further, this single principle (the Law of Contradiction) is sufficient for the whole of mathematics. . . . Because this is indisputably certain, we can easily be deceived into thinking that nothing more must be assumed for any other kind of knowledge."[54]

Kant, on the other hand, argues:

Accordingly, metaphysics has no formal or material grounds for certainty
different in kind from those of geometry. In both, the formal in judgment is
determined by the Laws of Identity and Contradiction. In both, there are
unprovable propositions which are the foundations for further conclusions.
. . .
   Since quantity constitutes the object of mathematics, and in considering
quantity we only pay attention to how many times something is given, it is
obvious that this knowledge must depend on a few and very clear first
principles of general mathematics (which is really general arithmetic).
. . . A few fundamental concepts of space permit the application of this
general mathematics to geometry. . . .
   Furthermore, there are only a few unprovable propositions at the basis of
mathematics; even if they were provable elsewhere, they are nevertheless
seen in this branch of knowledge as immediately certain: "The whole is
equal to all its parts taken together." "Between two points there can only
be one straight line." Etc. Mathematicians are used to setting such
principles at the basis of their discipline so that we become aware that no
other propositions than such obvious ones are taken as true while all the
rest are strictly proved.[55]

Thus it is quite clear that, following Crusius, Kant requires material principles in addition to the Law of Contradiction as a formal principle, but that
opposing Crusius, he also requires them for mathematics. In order to get a
general view of how Kant departed from Crusius, we can enumerate briefly the
ways in which principles are classified in the *Critique*.

   1. The Law of Contradiction as the supreme principle of all analytic
judgments.
   2. The supreme principle of all synthetic knowledge.
   3. The synthetic principles of pure intellect.
   4. The principles of geometry.
   5. The principles of arithmetic.
   6. The principles of the pure part of natural science.
   7. The analytic principles.
   8. The data of sensibility.

Of these, the Law of Contradiction and the data of sensibility conform to
Crusius's list. The supreme principle of all synthetic judgments and the principles of geometry and arithmetic are new. The analytic principles conform to the
lists of Leibniz, Wolff, and Crusius.
   In any case, it is clear that according to the Prize Essay of 1764, mathematics
also requires material principles. For geometry, these material principles coincide with the later synthetic axioms, as is shown by the examples (between two

points there can only be one straight line; space has three dimensions).[56] The question of which propositions Kant in the Prize Essay thought to be among the "few and very clear" principles of arithmetic is more difficult. Three groups of propositions come into question: (1) the whole is equal to its parts taken together (the later analytic principles);[57] (2) the axioms of arithmetic, the associative and commutative laws of addition; (3) the synthetic number formulas (e.g., according to B16, $7 + 5 = 12$). I am not able to decide with only the hints of the Prize Essay. Rather, we have to be satisfied with establishing that Kant in 1764, going beyond Crusius, required material principles for arithmetic as well as for geometry.

In Kant's time this relation to Crusius seems not to have been recognized right away in Königsberg. J. S. Beck is completely astonished to suddenly come across such a connection; he writes in a 1793 letter to Kant:

> A little while ago I read Crusius's *Weg zur Gewissheit und Zuverlässigkeit*, stimulated by Herr Schmidt's *Lexicon*, and to my surprise I've found (sec. 260) the distinction between analytic and synthetic judgments there much clearer than in the passage you cited from Locke. For although I think that he too shows no insight into the principles of synthetic a priori knowledge, still this passage contains enough so that it could well alert a thoughtful reader to its importance, because Crusius plainly indicates that this synthesis is the foundation of the reality of our concepts.[58]

Beck had thus read neither Kant's Prize Essay nor Crusius's *Weg zur Gewissheit* during his studies in Königsberg. It must be admitted that in substance Beck is right in his view that the systematic and historical link to Crusius of the account of synthetic judgment is stronger and more convincing than that to Locke.

In Kant's Dissertation of 1770, *On the Forms and Principles of the Sensible and Intelligible Worlds*, the previous twofold division of principles into formal and material is developed into a threefold one: (1) the material of sensible ideas, sensations (sections 4–5); (2) the principles of the forms of the world of the senses, space and time (sections 13–16); (3) first principles of the use of pure intellect, possibility, existence, necessity, substance, cause (section 8). The exposition of the principles is in conscious opposition to Wolff:

> But I fear that the illustrious Wolff, by his merely logical distinction between the sensible and the intellectual, has completely destroyed that noble debate of antiquity about *the nature of phenomena and noumena*, to the great detriment of philosophy, and has turned men's minds away from investigating them into investigating what are often merely logical minutiae.[59]

That Kant assumes synthetic axioms for geometry is shown by the almost complete agreement of section 15 with the corresponding passages about space in the *Critique*. In section 15, Kant gives a whole series of examples. The only question is whether he also establishes axioms for arithmetic. In section 12, mathematics is divided into geometry, mechanics, and arithmetic:

> Hence Pure Mathematics [*Mathesis Pura*] deals with *space* in Geometry [*Geometria*], and with *time* in pure Mechanics [*Mechanica pura*]. In addition to these concepts, there is another one which in itself is indeed intellectual but whose realization in the concrete requires the aid of the notions of time and space (by successively adding a number of things and by simultaneously putting them next to each other). This is the concept of *number*, which is dealt with in Arithmetic. So pure mathematics, which expresses the form of all our sensible knowledge, is the organon of all intuitive [*intuitivae*] and distinct knowledge. And since its objects themselves are not only the formal principles of all perception [*intuitus*],[60] b. ' are themselves *original perceptions* [*intuitus originarii*], it gives us the truest knowledge and at the same time an exemplar of what the greatest clarity is in other cases.[61]

The development of the first four chapters is briefly referred to at the beginning of the fifth chapter:

> *Practice gives rise to method* in all sciences whose principles are given in perception [*intuitive*], either by a sensible perception (experience) or at least by a perception which is sensible but pure (concepts of space, time, and number), that is, in natural science and mathematics.[62]

Particularly the beginning of section 23 shows quite clearly that principles derived from pure perception are also presupposed in arithmetic. This is certainly the approach to geometry and pure mechanics. Thus I cannot agree with Konrad Dieterich (1847–1888) in his 1877 *Kant und Newton* when he holds that Kant, in the Dissertation, thought of arithmetic as a purely logical field of knowledge, in contrast to geometry.[63]

The correspondence which Kant had with Lambert and Marcus Herz (1747–1803) about the Dissertation gives the same picture. Thus in 1770 Kant maintained that there were four kinds of principles of mathematics: (1) conceptual principles, (2) axioms of arithmetic, (3) axioms of geometry, (4) axioms of pure mechanics. Of these, especially the axioms of geometry are illustrated by many examples; but it is impossible to determine in the Dissertation which propositions Kant thought of as axioms of arithmetic.

In the 1790 *Streitschrift gegen Eberhard* Kant returns once more to the analytic principles: "For what can be proven merely philosophically from concepts, e.g., that the whole is greater than the part, does not belong to mathematics if its theory is established with complete rigor."[64]

Kant's characterization of the analytic principles here is the same as its development in the *Critique*. An analytic principle must be provable from concepts. When Kant asks how these propositions are to be classified, he faces an almost insurmountable difficulty. The propositions about equality, whole and part, etc. are difficult to order in one definite science and have always been shunted back and forth between ontology, logic, and mathematics. In the last few decades some propositions have, indeed, been dealt with quite fully in set theory and logistic; but it remains questionable whether the set-theoretical or logicist treatment is exhaustive. The question of where these propositions belong can even be posed with respect to Kant himself. He had a system of all the sciences in mind, at least as a goal. Where would the analytic principles be dealt with? No other place can really be found except in ontology, under the predicables of quantity. I have not, however, found any evidence of such an ordering in Kant, though he frequently speaks of the predicables yet to be treated, so that a reference to analytic principles must have been near at hand.

There are numerous Reflexions on the theme of the analytic principles. I can refer only to the most important. Four trains of thought can be distinguished which may also occur together in a single Reflexion:

1. Following Crusius, the Law of Contradiction as a purely formal principle is not enough; material principles are necessary.

2. There is terminological change from *material-formal* to *synthetic-analytic*.

3. How many kinds of principles are there?

4. If further material or synthetic principles are necessary, in addition to the formal Law of Contradiction and the material principles of sensation, what is the origin of these pure synthetic principles?

Reflexion 3709, from around 1762–63, leads to wholly original observations:

> Some knowledge must be the basis of other knowledge in our knowledge as a whole. . . . The basic concepts which do not presuppose still others are called *notiones primitivae* (primary basic concepts) and such kinds of judgment *judicia primitiva* (primary basic judgments).[65]

1. A series of Reflexions are related to the distinction Crusius had made between formal and material principles. Reflexion 3747, from around 1764–66: "All principles of human knowledge are either formal or material."[66]

Reflexion 3710, from around 1762–63:

All first principles are either formal or material. The first kind contains the ground of the way concepts in a judgment should be regarded in relations. The second contains the *medium terminum* by means of which they should be regarded in these relations with each other. . . . All propositions which come immediately under these two groups are unprovable; all which come mediately under them are provable.[67]

Reflexion 4655 is from 1771–78, according to Adickes, though purely from considerations of content this dating should be put back to about 1762–4.

The principles of identity and contradiction are both called the principle of contradiction. Indemonstrable judgments come under them directly. Material principles, in which the predicates are determinate, are indemonstrable. . . . The possibility of things calls for a real basis.[68]

These Reflexions again make the connection to Crusius clear, as Kant himself had indicated in his Prize Essay.

2. The second series of Reflexions presents the terminological transition from *material-formal* to *synthetic-analytic*.

Reflexion 3742, from around 1764–66, the second line from 1769: "There are material and also formal principles. Analytic principles, synthetic."[69]

Reflexion 3923, from around 1769: "Some principles are analytic and concern the formal in what is clear in our knowledge. Some are synthetic and concern the material."[70]

Some Reflexions use the new terminology throughout.

Reflexion 3750, from around 1764–66:

All first principles are either elementary propositions and analytic or are axiomatic and synthetic. Difference between an analytic and synthetic proposition in general. The rational are analytic, the empirical synthetic, the mathematical likewise.[71]

Reflexion 3744, from around 1764–66:

There are synthetic propositions from experience . . . they are also the axiomata of the mathematics of space. Rational principles cannot be synthetic at all. All empirical propositions are synthetic and the converse, all rational propositions are analytic.[72]

Reflexion 3976, from around 1769: "There are pure basic concepts of perception or reflection; the first are the principles of appearance, the second of insight. The first show coordination, the second subordination."[73]

Reflexion 3127, from around 1764–68: "The synthetic propositions increase knowledge materially, the analytic formally."[74]

Reflexion 3126, from around 1764–68, is also about this: "All principles are either formal or material."[75]

3. Two longer Reflexions from the years 1769 and 1771 concern themselves with the classification of the principles.

Reflexion 3923 (also quoted above in part):

Some principles are analytic and concern the formal about what is clear in our knowledge. Some are synthetic and concern the material, like the arithmetical, the geometrical, and the chronological; likewise the empirical. But there are still principles concerned with the use of reason in synthesis generally. Our reason, however, by its nature obeys the law that it does not know things directly but indirectly; thus it can only expect what happens to happen for a reason and finds irrational what is not determined by any reason.[76]

Reflexion 4370, from around 1771:

All immediately certain propositions are either
  1. basic formulas, or
  2. axiomata, or
  3. canons, or
  4. elementary propositions of analysis, or
  5. propositions of synthesis which are immediately certain.
The Laws of Identity and Contradiction are of the first kind. The second: objective principles of synthesis, space and time. The third: objective principles of qualitative synthesis. The fourth and fifth: material propositions contained immediately under the principles of the form and of analysis.[77]

These two Reflexions link the developments of both of the precritical writings referred to with the *Critique of Pure Reason* itself. They state directly which propositions should fall under which kinds of principles. The analytic principles are identified as elementary principles of analysis in Reflexion 4370, with the explanation "material propositions contained immediately under the principles of analysis." Reflexion 3923 speaks explicitly of arithmetic synthetic principles in conjunction with the geometric and chronological synthetic principles, which concern the material in knowledge. They thus explicitly confirm the interpretation of the Prize Essay and of the Dissertation just given. We have expressly decided that in both the 1764 and 1770 writings there are indications of arithmetical, geometrical, and chronological axioms. No statement of what Kant then thought to be among the axioms of arithmetic, or the pure sensible principles of numbers, can be gotten even from the Reflexions, and so no statement can be gotten from all the Kantian material now known.

4. The fourth group of Reflexions poses the question of the origin of those pure synthetic principles which were gradually elaborated.

Reflexion 4477, from around 1772–75:

Analytic propositions can be proved by means of the principles of contradiction or identity, but the synthetic ones cannot; so where do we get these from?
   1. empirically
   2. through pure perception
   3. through subjective conditions of the ideas of the intellect.[78]

Reflexion 4162, from 1769–70:

All knowledge has either empirical or rational principles; the latter are either logical or real. The logical take the primary concepts from things and relations as they are given and regard only the subsumption under identity and contradiction; the second are primary concepts and relations given through the nature of reason.[79]

That this became the decisive question for Kant is shown again in his long letter to Marcus Herz.[80] If, as it seems, Crusius distinguishes formal and material principles, and if he thereby only interprets the data of sensation (as Kant interprets them) as falling under material principles, such a question offers no basic difficulties. In any case, the givenness of sense data is not a problem in this regard. The question only becomes urgent when Kant has to stick material but pure principles in between formal principles and material sense data. When these pure material principles are traced back to a pure perception, he is only giving a preliminary sketch. The letter to Herz and the two Reflexions 4477 and 4162 not only show the clarity with which Kant worked out the problem, but they also show his original answer: the pure synthetic principles arise from the nature of human reason.

All Kantians (in the narrower sense of his own students) agreed on the question of the analytic principles and also on the interpretation of the propositions as well as on what concerns the set of circumstances here to be considered.

Kiesewetter put an introduction about the object and methods of mathematics at the beginning of his *Die ersten Anfangsgründe die reinen Mathematik* of 1799. In section 9, he gives this table of analytic principles:

   1. When two quantities are equal to a third, they are equal to each other.
   2. When two quantities are similar to a third, they are similar to each other.
   3. When two quantities are congruent to a third, they are congruent to each other.

4. Every quantity is equal to itself.
5. The whole is greater than a part of itself.
6. The whole is equal to all of its parts taken together.[81]

Zimmermann also gives a list of ten analytic principles in his *Entwicklung analytischer Grundsätze für den ersten Unterricht in der Mathematik* of 1805:

1. Every quantity is equal to itself.
2. Two quantities equal to a third are equal to each other.
3. Equals added to equals give equal sums.[82]

Principles 4 through 10 modify principle 3 for the other operations, namely, subtraction, multiplication, division, raising to powers, and inequality.

The first metaphysics written in the Kantian manner, *Grundriss der allgemeinen Logik und kritischen Anfangsgründe zu einer allgemeinen Metaphysik* by Ludwig Heinrich Jakob (1759–1827) in 1788, deals with some of the analytic principles in the part entitled "Of the Pure Concepts" in the first chapter "Of Predicates and Principles of Quantity."[83] Thus the analytic principles in a Kantian system of knowledge are dealt with in systematic ontology as the predicables of quantity.

It is generally recognized that Kant emphasized a particular mathematical state of affairs by his distinction between formal and material principles, or between synthetic axioms and analytic principles. To be sure, it is noteworthy that the problems concerning the analytic principles have really only aroused the interest of mathematicians while the purely philosophical interpretations have either missed the point or have just ignored the problems. Georg Simon Klügel (1739–1812), who was sympathetic to Kantian ways of thinking, accepted the Kantian body of ideas. He had this to say in his *Mathematisches Wörterbuch* of 1803:

> A principle is a proposition which in itself is so obvious that it need no proof. The principles can be classified into logical or formal and material. The former include a form of reasoning peculiar to mathematics. The first seven principles in the first book of Euclid's Elements are included in them. What is equal to one and the same thing is equal to itself, equals added to equals give equals. . . . The material principles flow directly from the concepts which they are concerned with. What includes and is included by another is equal to it, the whole is greater than the part, all right angles are equal to each other, two straight lines do not enclose a space.[84]

The significance of the analytic principles was then noticed by Frege. Hankel dealt with them in great detail in his *Theorie der komplexen Zahlensystem* of 1867:

One would think that even superficially the distinction could be made between two essentially different classes of principles, of which one . . . refers to relations basically connected to the concept of equality while the other . . . contains geometrical truths. And yet this distinction has been completely overlooked by most mathematicians. This is sufficiently shown by the fact that both have been thrown under the name of axiom, which Euclid did not use at all because he had recognized this distinction with the greatest clarity. . . . There can be no doubt that the principles stemming from geometrical perception are synthetic . . . in Kant's sense. . . . We are more familiar with the other class of principles, that of the *notiones communes*, which Kant had indeed recognized as different from the former. . . . Such a principle expresses an abstract and necessary law found in all fields of quantity which, without giving up its essential character, can be changed into a definition. Furthermore, there is such a high degree of evidence for it that it can be recognized as indubitably true merely by an exposition of it.[85]

Günther Thiele (1841–1910) in his 1869 Dissertation *Wie sind die synthetischen Urteile der Mathematik a priori möglich?* is surely wrong when he interprets the passage at B17 as indicating some uncertainty on Kant's part.[86] We have already shown above that one can properly ask about some of these propositions whether the classification as analytic principles is justified or whether they must not instead be regarded as synthetic axioms. This question, however, has to be considered from entirely different bases. Thiele starts here from much too narrow a conception of analytic judgment; if *analytic judgment* is mistakenly taken in so narrow a sense, there will hardly be any such judgments.

Kant recognized a whole series of propositions at the basis of mathematics which he identified as analytic principles. Complete compilations are to be found in the textbooks of his students, particularly in Schultz's *Anfangsgründe der reinen Mathesis*. The Kantian analytic principles include the Euclidean principles 1, 2, 3, 8, according to Heiberg's edition of Euclid's *Elements*, and 1, 3, 4, 5, 6, 7, 9, 19, according to Clavius's edition.[87] Of these the following are to be found in Kant's works:

The whole is equal to itself: CPR, B17; *Prolegomena*, 169; Reflexion 4634.

Equals added to equals give equals: CPR, A164, B204.

Equals from equals give equals: CPR, A164, B204.

The whole is greater than the part: CPR, B17; *Prolegomena*, 269; Reflexion 3998; *Streitschrift gegen Eberhard*, Ak. VIII, 196.

The whole is equal to all of its parts taken together: Prize Essay, Ak. II, 281.

Three groups of propositions are called into question as analytic propositions: (1) completely empty propositions like "The line is a line"; (2) the analytic principles; (3) propositions which follow directly from definitions such as "A square has four sides." If the totally empty propositions arising only from pointless formalization are left out, then both objectively and historically we can best say that for Kant the analytic judgment is described thus: Analytic judgment can be proven, in the sense of the Leibniz-Wolff theory of judgment, by definitions together with the laws of identity and contradiction. It follows directly that an analytic judgment for Kant must be a priori. Furthermore, it must be possible to derive the judgment from the definition, that is, from the concept of the subject, and thus we find the link to the true description. We may then conclude: (1) Each of the analytic principles is to be found in the works of Kant and his students. (2) It is clear how analytic principles are proved—from definitions and the law of contradiction. (3) The link to Wolff and Leibniz is established.

CHAPTER THREE

The Axioms of Arithmetic

T HE AXIOMS OF ADDITION were published by Johann Schultz in 1789–92 in the *Prüfung der Kantischen Critik der Reinen Vernunft*, in 1790 in the *Anfangsgründe der reinen Mathesis*, and again in 1805 in *Kurzer Lehrbegriff der Mathematik* (2d ed., 1820). From then on they appeared in numerous works whose authors can be called Kantians in either a narrower or a wider sense. Kiesewetter and Zimmermann were actually students of Kant's. The first statement of the axioms given from outside the small circle of students occurs in *System der Elemente der allgemeinen Grössenlehre* by Friedrich Wilhelm August Murhard (1729–1853) in 1798. Martin Ohm (1792–1872) and Jakob Fries then followed. The axioms were taken over into mathematics proper by Hermann Grassmann, Sir William Rowan Hamilton (1805–1865) and Hermann Hankel. In this way the axioms of arithmetic gradually appeared in specifically mathematical textbooks, for example, those of Ernst Schröder (1841–1902), Gottlob Frege, Hermann Schubert (1848–1911), Otto Stolz (1842–1905), Richard Baltzer (1818–1887), and Carl Anton Bretschneider (1808–1878). Peano gave the final statement in 1889 in his *Arithmetices principia*.

After giving an account of these two groups, I will try to show that there is a direct line from Kant to modern mathematics, basically through Ohm and Grassmann, although it is not really possible to establish this conclusively in view of the great number of works of the Kantians in which the axioms had already appeared. At the end of this chapter I will examine the question of whether the axioms stem from Kant or Schultz.

To show unequivocally Kant's importance as a productive mathematician I take the commutative law of multiplication as an example, so that the originality and importance of Kant's procedure will become clear.

First, here is Schultz's account in the *Anfangsgründe,* which is dated 1790 (although reference is made in it to the *Prüfung*, which bears the date 1791 on its title page.) He says:

Axioms and postulates of general combinations of quantities.

Postulate 1: To transform many given homogeneous quanta into one quantum, i.e., into a whole, by taking them together successively.

Note: That and how this is universally possible is immediately and intuitively evident, without a rule being given for it, so that this proposition is a postulate (*Proleg.* sec. 33). This is shown more fully in my *Prüfung der Kantischen Critik,* pp. 221ff.

Postulate 2: To increase and decrease in thought any given quantum without end.

Note: That and how this is possible is immediately evident without anyone being in the position to give any further rule for always being able to think of a greater or smaller, no matter how great or small a given quantity may be. There is more about this in my *Prüfung der Kantischen Critik,* pp. 223, 224.

Axiom 1: The quantity of a sum is one and the same, whether the first given quantum be added to the other one or the other one to the first, i.e., it is always the case that $a + b = b + a$, e.g., $5 + 3 = 3 + 5$.

Axiom 2: The quantity of a sum is one and the same whether to a given quantum another be added either as a whole or by each of its parts, one by one, i.e., it is always the case that $c + (a + b) = (c + a) + b = c + a + b$.[1]

Then there are propositions of interest to us to be found in different places in Leibniz's works. The law of the unlimited feasibility of addition occurs in the remarks about the first book of Euclid,[2] while the commutative law appears in his "Prima calculi magnitudinium elementa demonstrata in additione et subtractione, usuque pro ipsis signorum + et − ":

(8) Theorem: $+ a + b = + b + a$.

It is clear from the preceding that it does not matter in what order they are put together: it is sufficient that one is put together with another.[3]

The commutative and associative laws are then repeated in the *Specimen Calculi universalis* in the manuscripts published by Carl Immanuel Gerhardt (1816–1899) and Louis Couturat. I have not, however, been able to find any evidence that these two laws were made public in any way; in any case, they do not occur in the textbooks of the eighteenth century, certainly not in Wolff's.

When Schultz says that arithmetic has principles "no matter how much they have been overlooked up till now," in the passage where he himself calls attention to it, he quite consciously wishes to claim priority in discovering these principles.[4] His claim is true so far as the two axioms go, but the two postulates had already been dealt with repeatedly. The second postulate expresses the second postulate of the seventh book of Clavius's edition of Euclid's *Elements*:

"Given any number, there is always a larger number." Wolff speaks in just the same way about the same postulate in the *Ontologia*:

> When any number is given, a larger one can be given. For let any number be given: since numbers are produced by the continual adding of unity, there is nothing to prevent you from adding unity to the given number, either one or several times, that is, from adding any other number.[5]

However, the proposition was recognized as a postulate by Kant while for Wolff it was a theorem. Lambert's treatment of it in his *Neues Organon* of 1764 is different for he describes the unlimited increase by addition of 1 as a postulate:

> The concept of unity is likewise simple, and we have it immediately in the word *me* and so also in the presentation of every concept, insofar as it is a concept. The repetition of the unit gives us the concept of number which is the object of arithmetic and thus is knowledge a priori because, except for the concept of the possibility of this repetition, no other additional postulate is necessary.[6]

A passage where Lambert practically speaks of two postulates contradicts this in a way:

> We have the concept of unity in the word *me* and, indirectly, in that which we hold together in our ideas. The repetition of the unit gives the concept of number and is in itself a postulate which the whole of arithmetic depends on. We have already noted all this in sec. 26 and, accordingly, add only that this taking together of many units makes a number which we can, in any case, regard as a unit. That this could go on as long as desired is held to be among the postulates.[7]

The postulate here, that combining many units shall constitute a number, corresponds to the Kantian Postulate 1.

Kiesewetter presents the principles of arithmetic in his 1799 *Die ersten Anfangsgründe der reinen Mathesis*, as in all his works, in popular form:

> The ones are added when the units of one number are added to the other one by one. It is a postulate of arithmetic that this can be done. . . . It is just the same whether the first number is added to the second or the second to the first, $5 + 4 = 4 + 5$.[8]

Thus Kiesewetter knows the propositions. The organization of his textbook is so flabby, however, that the axioms and postulates are not distinguished from the theorems in either arithmetic or geometry, although the introduction deals with such a distinction in detail.

An unaltered account occurs in the work of Christian Gottlieb Zimmermann, a favorite student of both Kant and Schultz who studied mathematics and philosophy in Königsberg from 1788 to 1790.[9]

Another account, not independent of these, appears in Friedrich Murhard's *System der Elemente der allgemeinen Grössenlehre*.[10] He has taken over the axioms and postulates word for word [from Schultz's *Anfangsgründe*] in his *Grössenlehre*, apart from quite unimportant changes (such as *Grösse* [quantity] for *Quantität* in the first postulate):

> Arithmetic has postulates as well as axioms. The axioms are:
> 1. The quantity of the sum is one and the same, whether the first given quantum be added to the second or the second to the first, i.e., it is always the case that $A + B = B + A$.
> 2. The quantity of the sum is one and the same, whether to a given quantum another be added either as a whole or by each of its parts, one by one, i.e., it is always the case that $C + (A + B) = (C + A) + B$. The postulates, however, are:
> 1. To derive the concept of a quantity from many given homogeneous quantities, i.e., to transform them into a whole, by taking them together successively.
> 2. To enlarge and decrease a given quantum as much as we want, i.e., into infinity.[11]

A 1799 letter from Gauss to Wolfgang Bolyai refers to this *Grössenlehre* of Murhard's.

> I wish your emperor's land luck in having acquired Murhard. The book which he has dedicated to the emperor is word for word (I say word for word) plagiarized from three others (Schulze, Segner, Stahl), apart from a few passages which are beneath contempt. (I have this information from Pfaff, and I have myself made comparisons and found word for word correspondence.)[12]

The editors of the Gauss-Bolyai *Briefwechsel* mistakenly think that Schulze is Johann Karl Schulze (1749–90), who had a short reference book on geometry, *Taschenbuch der Messkunst*, published in Berlin in 1782–83.[13] Their mistake arises solely from the fact that Murhard's *System der Elemente der allgemeinen Grössenlehre*, the book in question here, was not available to the editors because of its rarity, as they explain. If we look at Murhard's work first hand, we see right away that Murhard copied from the *Anfangsgründe*[14] of the Königsberg mathematician. I will compare short passages with Schultz's *Prüfung* which are even more interesting because Schultz for his part himself took the passages from a letter which he had received from Kant.

1. Original: Kant's letter to Schultz, 25 November 1788, Ak. X, 555:

General arithmetic (algebra) is a science which is expanding[15] so much that no other rational science can be named its equal in this respect.

2. Schultz, *Prüfung*, 1789, 232:

Now general mathematics [*Mathesis*] is a science which is capable of such astonishing expansion merely a priori that no other rational science can be named its equal in this respect.

3. Murhard, *System der Elemente der allgemeinen Grössenlehre*, 1798, 39:

Furthermore, general mathematics [*Grössenlehre*] is a science which is capable of such astonishing expansion a priori that no other rational science can be named its equal in this respect.

Kant:

even, that the development of the remaining parts of pure mathematics [*Mathesis*] depends for the most part on the expansion of general mathematics.

Schultz:

Hence the development of the whole of special mathematics [*Mathesis*] depends for the most part only on the expansion of the general.

Murhard:

Hence the development of the whole of special mathematics [*Grössenlehre*] depends for the most part only on the expansion of the general.

Kant:

now if these consist of merely analytic judgments then, at the very least, the definition of them as merely explanatory judgments would be wrong, and then there would be an important and difficult problem of how the expansion of knowledge is possible by merely analytic judgments.

Schultz:

Thus it is an obvious contradiction for a science of such enormous scope to arise merely by analyzing the concepts of unity, plurality, and totality; so this fact alone shows that it must be synthetic throughout.

Murhard:

> It is thus an obvious contradiction for a science of such enormous scope to arise merely by analyzing the concepts of unity, plurality, totality. So this fact alone also shows that it must be synthetic throughout.[16]

Schultz makes use of Kant's letter as Kant intended him to. The letter from Gauss to Bolyai is interesting in many respects. It shows that Schultz was so well known that he could be cited without further explanation in a letter between two friends, and that Murhard's plagiarism was noticed right away. Other citations also indicate that Schultz's works were well known. Here I will only cite Friedrich Novalis (1772–1801),[17] and Martin Ohm's *Kritische Beleuchtungen der Mathematik*[18] of 1819. Gauss's letter to Bolyai shows, moreover, that in 1799 Gauss knew of the axioms of arithmetic and their origin.

By 1800, therefore, the axioms of arithmetic had already appeared in three works (by Kiesewetter, Murhard, and Schultz) under the direct influence of Kant. They then got into the textbooks of the nineteenth century, in part in quite a woolly form. First there is Bernhard Friedrich Thibaut (1775–1832), who worked with Gauss in Göttingen. Fries, Ohm, and Grassmann really continued the work with the axioms of arithmetic further. Jakob Fries put his mathematical work in his *Die mathematische Naturphilosophie* of 1822. The division of subjects in this work shows that Fries was basically interested in arithmetic and combinatorics, while geometry was secondary. This work is the first attempt at a systematic axiomatics; the axioms were supposed to be derived from a single basic principle:

> The supreme form of all axioms is:
> It is always the case that between given boundaries, one and only one series is possible.
> The forms of the postulates are as follows:
> 1. Postulates of the description of a quantity.
> To describe a quantity geometrically by a geometrical motion in the imagination—to construct a quantity arithmetically by the imagination which puts together homogeneous parts.
> 2. Postulates of delimiting.
> To delimit parts of every continuous quantity in imagination.
> 3. Postulates of enlargement.
> In thought to increase every imagined quantity.
> 4. Postulates of ordering.
> In thought to vary every given ordering.
> 1. Axiom. From homogeneous quantities taken as parts, one and only one sum is always possible.
> Addendum 3. A quantity as a whole can always be set as equal to any

arbitrary combination of all its parts for only one quantity is possible from these parts.

All complexes from the same elements are equal to each other.

2. Requirements:

1. In thought to add a unit to every given quantity.

2. In thought to subtract a unit from every given quantity.[19]

The axiom of arithmetic very ingeniously contains three of the Kantian principles. It is explicitly made clear in Addendum 3 that the commutative and associative laws are contained in it. The first Kantian postulate is formulated in the axiom "One and only one sum is always possible." The second postulate is subdivided into two requirements, since Fries already knew the principle of the unlimited feasibility of inverse operations—which is generally first ascribed to Hankel. But it is difficult to regard Fries's treatment of Kantian thought as an advance. First, the work suffers from thinking too much in terms of a system; up till now at least, all attempts at a systematic mathematics have been sterile. Then too, the derivation of addition from combination cannot be convincing. The clarity with which Kant threw the basic laws into relief is, moreover, pretty well obliterated.

Yet Jakob Fries is not only familiar with the individual axioms but is also clear about the significance of axiomatics, his systematizing being really of value only as an attempt. It has to be admitted that this systematizing goes directly back to Schultz, whose work Fries cites often and the knowledge of whom he presupposes in his readers, as is shown by his citations in *Die mathematische Naturphilosophie*.[20] (Fries's dependence on Kant and Schultz, assumed from the outset, is actually shown by these citations.)

Martin Ohm presents the four principles in the following way in his *Versuch eines vollkommen consequenten Systems der Mathematik* of 1828–32:

Addendum: Given the numbers a and b, then $a + b$ also always signifies a number, but only a unique one.

3. It is always the case that $a + b = b + a$.

For in both cases we have the number which has as many units as the numbers a and b together.

Theorem 1. A number c is added to a sum $a + b$ when this number c is added to one of the terms of the sum, while the other term is left unchanged, i.e., $(a + b) + c = (a + c) + b = a + (b + c)$.[21]

Both postulates are combined here into one—addition always gives a unique number; both of the other propositions express the two axioms. All four principles are presented as proven propositions and, indeed, for Ohm the proof follows from his definition of addition: "We represent the number $a + b$, which consists of the two terms a and b, as having as many units as the numbers a and b taken together."[22]

The account in the *Elementar-Zahlenlehre* of 1816 is similar:

(Def. p. 7).
The number c, which is composed of as many units as the two other numbers a and b taken together, is denoted by the representation of $a + b$, and this representation is called the sum of the two numbers a and b.[23]

With this definition, Ohm proves eight theorems, including the following:

5. The representation of $b + a$ denotes the same number as the representation of $a + b$, i.e., $a + b = b + a$.
6. It is the case that $(a + b) + c = (a + c) + b = a + (b + c)$, etc.; i.e., the representations of $(a + b) + c$, $(a + c) + b$, $a + (b + c)$, etc., all denote the same number, namely, the number which has as many units as the numbers denoted by a, b, c taken together.[24]

Note that the precision of Schultz's formulation of the reason for the unprovability has been entirely lost here; the question is not at all whether something *may* be done—say whether the sides may be interchanged for each other. Why should a mathematician be prevented from interchanging the two terms in a sum? The interchange is always possible and the question is rather whether the sum remains unchanged afterward. It can hardly be expressed better than Schultz does when he says, "But how do I know that this arbitrary procedure, which is not part of the concept of addition itself at all, [. . .] does not alter the quantity of the sum of $7 + 5$?"[25]

Thus, while Ohm took over the content of the axioms, he lost the fineness of the structure, and the system-building becomes intolerable. (Incidentally, Ohm's dependence on Kant and Schultz is obvious when one compares the introduction of Ohm's *System der Mathematik* with the introduction to Schultz's *Anfangsgründe*.)

Hermann Grassmann gives the decisive presentation in his *Lehrbuch der Arithmetik*. He denotes the positive unit term e:[26]

[8–9 Explanation]: If a is any one of the members of the basic series, we define $a + e$ . . . to be the member of the series immediately following a.

. . .

[15] Explanation: When a and b are any arbitrary members of the basic series, we define $a + e$ to be the member of the series for which the formula

$$a + (b + e) = a + b + e$$

holds. a and b are called the terms or parts of the sum $a + b$, a the first term, b the second. The joining is called addition. . . .

20. $e + a = a + e$.

Proof (relative to a): Assuming that the formula 20 holds for any quantity a, I will show, first of all, that it also holds for the quantity $a + e$ immediately following a, i.e., that

$$e + (a + e) = a + e + e.$$

We have

$$e + (a + e) = e + a + e$$
$$= a + e + e$$

since, by assumption, formula 20 should hold for the given value a. . . .

Now, however, formula 20 holds for case $a = e$, for then

$$e + a = e + e = a + e^{27}$$

Later developments have shown that this derivation of Grassmann's cannot be improved upon. Grassmann, incidentally, makes things complicated for himself by making the derivation for the set of all the whole numbers at once, i.e., for negative as well as positive numbers. He comes somewhat too quickly to the question of whether addition is always possible, the question of the unambiguity of addition, in particular, the meaning of the inference of n to $n + 1$ for the whole numbers. (These three factors are dealt with acutely in Peano's definitive work.)

Whether the formula $a + (b + e) = a + b + e$ is a definition or an axiom may be a question of the ontological interpretation of mathematics. The answer to the specifically mathematical question is completed by bringing out the minimal assumptions clearly. Grassmann himself holds the view that arithmetic consists only of definitions and thus has no axioms:

> Proof in the formal sciences, therefore, does not go beyond the thinking itself to another sphere but remains purely in the combining of the different acts of thought. Thus these formal sciences cannot proceed from axioms like real knowledge; rather, definitions form its basis. When axioms are, nevertheless, introduced into formal science, as for example into arithmetic, this is to be regarded as an abuse which can only be explained by reference to the related treatment of geometry.[28]

Sir William Rowan Hamilton knew of all five laws but gave no systematic exposition of the arithmetic of whole numbers.

Hermann Hankel starts with a general theory of form in section 2 of his 1867 *Theorie der komplexen Zahlensysteme*. He deals there, in section 4, with the algorithm of associative operations without commutation, in section 5 with the algorithm of associative operations with commutation. An arithmetic having five axioms is dealt with in order to discover what can be derived from the individual laws or from the combinations of some of these laws respectively. He investigates the validity of these laws in section 3:

> The formal laws of addition and multiplication established in the last two paragraphs are the familiar laws introduced in secs. 1, 2 of actual addition and multiplication in arithmetic, reproduced with some freedom (relative to commutativity). These are the laws which we will carry over into the realm of perception, in particular of space, in what follows, and this is one aspect of the principle of permanence of the formal laws.
>
> We will thereby in general proceed thus: When a field of objects is given one will first of all ask whether there is an operation applicable to it which has the properties of addition. There is no exact method for answering this question, however; rather, creative discovery must solve it. The principle of permanence stands us here in good stead. If, however, an operation is found which has the properties of addition, one will ask whether there is a corresponding multiplication.[29]

We can sum up Hankel's point of view by saying that there is a purely logicist general theory of form in itself, but that an axiom system must be established for every "field of objects" for which possible laws developed in the theory of form are valid. The numbers, especially the real numbers, constitute such a realm for Hankel. The systematic exposition of the addition and multiplication of the positive whole numbers is on 37–40. Although Hankel expressly refers to Grassmann, saying, "The idea of deriving the rules of addition and multiplication, as it appears here, is basically due to Grassmann (*Lehrb. d. Arithm.*),"[30] it seems to me that Hankel makes progress since the development is not undertaken at the outset for all whole numbers but remains limited to the positive whole numbers. As before, in Hankel there is still no stress on the inference from n to $n + 1$ or on the Archimedean axiom.[31]

Gottlob Frege's work is primarily in his 1884 *Foundations of Arithmetic* in which he raises the following strong objection to Grassmann's system:

> (Grassmann) wants to derive the law
>
> $$a + (b + 1) = (a + b) + 1$$
>
> by definition when he says: "When a and b are any arbitrary members of the basic series, we define the sum $a + b$ to be the member of the series for which the formula
>
> $$a + (b + e) = a + b + e$$
>
> holds."

Thereby *e* means the positive unit. This explanation can be criticized in two ways. First of all, the sum is explained in terms of itself. If one does not yet know what $a + b$ means, one does not understand $a + (b + e)$ either. This objection may, however, be overcome by saying (contradicting the actual words, of course) that it is not the sum but the addition which is to be defined. Then it would still be possible to object that $a + b$ would be a meaningless symbol if there were no member of the basic series or if there were not more than one of the required kind. Grassmann simply assumes that this does not occur and does not prove it, so that the rigor is only an apparent one.[32]

The basic laws were presented definitively by Giuseppe Peano in the work which appeared in 1889, *Arithmetices Principia*. In merely twenty pages this work provides an abundance of arithmetical laws in an extraordinarily clear and elegant style.

Peano begins with new axioms. Axiom 1 establishes the concept of whole number. Axioms 2, 3, 4, and 5 are the axioms of equality. Axioms 6 and 7 establish that addition is always possible and gives unique results. Axiom 8 is new and important; it precludes adding in a circle. Axiom 9 finally stipulates the validity of the inference from n to $n + 1$ for the set of the whole numbers, which is what he is examining. Peano then proves the basic laws of addition with these new axioms. Proposition 18 defines the general concept of addition in the way proposed by Grassmann:

$$a + (b + 1) = (a + b) + 1.$$

Proposition 23 presents the associative law, which holds, first of all, for $c = 1$ because of definition 18. But if it holds for c, it also holds for $c + 1$. Thus, it holds in general.

The proof of the commutative law follows in two steps. First, in theorem 24 the proposition is proved for the special case that one term of the sum $= 1$ so that $1 + a = a + 1$. Then the general proof that $a + b = b + a$ is given in proposition 25. Both proofs are derived by means of the inference from n to $n + 1$.[33]

It is quite clear from this that the definition first used by Grassmann is a special case of the two laws from which both laws can be derived in complete generality with the help of the inference from n to $n + 1$. This is immediately clear for the associative law. It becomes clear for the commutative law when a and b are set $= 1$. For then from

$$a + (b + 1) = (a + b) + 1$$

we get

$$1 + (1 + 1) = (1 + 1) + 1$$

or

$$1 + 2 = 2 + 1$$

After the connection between the basic laws is finally established by this argument, the ontological questions can be taken up. What do *axiom* and *definition* mean here? For example, can definition 18 be stipulated so simply? Frege had already criticized such a claim. The more recent discussions—of Hilbert, Natorp, Couturat, Zermelo, Russell—deal almost exclusively with the ontological questions, which we can now put in this way: Is arithmetic logistically or axiomatically based?

I now want to discuss the commutative law of multiplication in order to eliminate any doubt that all the formal precision of modern mathematics had already been required by Kant.

The commutative law of multiplication, $a \times b = b \times a$, was discovered by Blaise Pascal (1623–1662), according to Cantor.[34] Schultz states the proposition and the proof in the following way in the *Anfangsgrunde*:

A product of whole numbers is the same whether the first factor be multiplied by the second or the second by the first, i.e., $n \times r = r \times n$.

For since $n = 1$ times n (sec. 36, note 3) $= n$ times 1 (sec. 36, note 2), therefore $n \times 1 = 1 \times n$.

$$n \times 2 = n \times 1 + n \times 1 \text{ (note 6)} = 1 \times n + 1 \times n \text{ (p. dem.)}$$
$$= 2 \times n \text{ (note 4)}$$
$$n \times 3 = n \times 2 + n \times 1 \text{ (note 6)} = 2 \times n + 1 \times n \text{ (p. dem.)}$$
$$= 3 \times n \text{ (note 4)}$$
$$n \times 4 = n \times 3 + n \times 1 \text{ (note 6)} = 3 \times n + 1 \times n \text{ (p. dem.)}$$
$$= 4 \times n \text{ (note 4)}[35]$$

Thus Schultz had already recognized that the commutative law of multiplication can be proved by means of the inference from n to $n + 1$ from the special case $1 \times n = n \times 1$ and by the distributive law:

$$(a + b) \times n = a \times n + b \times n = n \times a + n \times b.$$

In contrast, Martin Ohm completely falls back on the old geometrical schema:

It is always the case that $a \times b = b \times a$, and because of this, a and b are both called, without distinction, factors of the product. For imagine a number b of horizontal rows and in each of these horizontal rows a units, as is to be seen in the following schema.[36]

The same is true of the proofs of Adrien Marie Legendre (1752–1833) and Joseph Diez Gergonne (1771–1859). Grassman's, on the other hand, is completely identical with the Kantian proof:

If a and b are numbers, then

$$ab = ba$$

The factors of a product can be interchanged if they are numbers. Proof (by induction). The formula 72 holds for any numerical value b, so $a(b + 1) = ab + a$

$$= ba + a$$
$$= ba + 1a$$
$$= (b + 1)a$$

Now formula 72 holds for $b = 1$; for

$$a \times 1 = a = 1 \times a$$

hence etc.[37]

We will now show the actual connection between Kant and these specifically mathematical discussions. I have to say that I was not able to determine whether Sir William Rowan Hamilton also goes directly back to the Kantian work; I did not have the historical material to pursue the development in England or France. Hamilton himself gives us a helpful hint in the preface to his 1853 *Lectures on Quaternions*:

> I was encouraged to entertain and publish this view, by remembering some passages in Kant's *Criticism of the Pure Reason*, which appeared to justify the expectation that it should be *possible* to construct, a priori, a Science of Time as well as a Science of Space.[38]

To be sure, Hamilton here speaks only of being encouraged, so that no positive conclusion can be drawn. What is significant for us is that the creator of the theory of quaternions, which is formal, sees his intentions as being encouraged by Kant, a strong objection to the usual interpretation of Kant's mathematics as perceptual.[39]

Ohm is cited by Hamilton in many passages, among others the following:

> It is proper to mention, that results substantially the same, respecting the entrance of two arbitrary whole numbers into the general form of a logarithm, are given by Ohm, in the second volume of his valuable work, entitled: *Versuch eines vollkommen konsequenten Systems der Mathematik, vom Professor Dr Martin Ohm* (Berlin, 1829, Second Edition, page 440. I have not seen the first edition).[40]

On the other hand, the German development can undeniably be traced back to Grassmann and Ohm; Peano expressly refers to Grassmann: "In demonstrations of arithmetic I use the book: H. Grassmann, *Lehrbuch der Arithmetik*, Berlin, 1861.[41]

Hankel refers to Grassmann and Ohm in the same way in 1867 in the *Theorie der komplexen Zahlensysteme*:

> After I had recognized that the method I had adopted in this work was the only scientifically satisfactory one, I applied myself to finding out how far it had already been pointed out by others. I have not gotten very much. . . .
>
> In Germany, M. Ohm presented a purely formal exposition of arithmetic operations in the first edition of his *Versuch eines vollkommen konsequenten Systems der Mathematik* in 1822; he then altered it in the second edition of 1828 for the sake of greater popularity without changing his view that "only this way is logically rigorous and thus it alone guarantees complete certainty." The result was that he started from the usual concept of number and always applied it in combination with that of a formal operation.
>
> Ohm's ideas appeared here and there in German textbooks in a similar though even less rigorous way but without the ideas having been developed so consistently, so far as I know, as has probably been done by the English. . . .
>
> H. Grassmann was the first to grasp these ideas in a truly philosophical spirit and to consider them from a comprehensive point of view.[42]

Thus the development is indisputably traced back to Ohm and Grassmann. The following note from the biography of Grassmann by Friedrich Engel (1861–1941) gives information about the relation between Grassmann and Ohm:

> It is really remarkable how few mathematical books Grassmann got along with during his whole life. His mathematics library was extremely small since he mainly used the relatively large library of the Gymnasium where, among others, the *Crelle'sche Journal*, Poggendorff's *Annalen der Physik und Chemie*, and also the *Mathematischen Annalen* and even the *Comptes Rendues* were available to him. Only the following works about higher mathematics can be identified as having been in his private possession: Lagrange, *Théorie des fonctions* and *Mécanique analytique*, Poncelet, *Traité des propriétés projectives*, Moigno, *Calcul différentiel et intégral*, Verhulst, *Traité des fonctions elliptiques*, Ohm, *System der Mathematik*, Magnus, *Aufgaben aus der analytischen Geometrie*, Steiner, *Systematische Entwicklung* and *Die geometrischen Konstruktionen, ausgeführt mittels der gerade Linie und eines festen Kreises*, Möbius, *Barycentrischer Kalkul*, and later, most likely a present from the author, *Mechanik des Himmels*.[43]

We now ask whether the principles of arithmetic stem from Kant or from Schultz. To begin with, at the time it was not only generally believed that Schultz was the only genuine interpreter of Kant, but this is also expressly stated by Kant himself.

Furthermore, Wilhelm Dilthey (1833–1911) showed that Schultz, for his review of the 1790 essays of Abraham Kästner (1719–1800) in J. A. Eberhard's *Philosophisches Magazin*, used a manuscript made available to him by Kant for this purpose, practically word for word.[44] Schultz also sent the manuscript of his first commentary, the 1785 *Erläuterungen über des Herrn Professor Kant Critik der reinen Vernunft*, to Kant for examination before publication, even though Kant apparently took no direct interest in this first work.[45]

It can be shown from the *Prüfung der Kantischen Critik der reinen Vernunft*, where the principles of addition first appeared, that the parts relating to arithmetic basically go back to Kant. Schultz had also sent this work to Kant for prior examination. The judgments of arithmetic were identified as analytic judgments in the manuscript. This was in accordance with Schultz's own point of view. Shortly before in 1786, he had only defended the thesis "Judgments of geometry are synthetic" in his inaugural lecture for his professorship of mathematics.[46]

Such a restriction to geometry was unambiguous, for the *Prolegomena* had appeared in 1783 and had presented a full account of the synthetic character of the judgment $7 + 5 = 12$. Kant then offered Schultz a second examination of this question in a detailed 1788 letter; at the conclusion of the letter he invited him to talk over this question in person:

> To my way of thinking, "honesty is the best policy" is a motto for everything, including writings which are concerned with the improvement of human knowledge and especially with the honest, candid examination of our faculties, not to deceive by the concealing of the mistakes we find in our own system or by partisanship and contention for the sake of winning. Thus, I only wanted to have a look at the painstaking work you are now starting on before it is published in order to make it easier to eliminate a misunderstanding, which is easily done by a mutual exchange of ideas (which is so easy here since we are so near each other) and so to forestall many future controversies.
>
> Allow me, therefore, to bring up some difficulties with the claim which is opposed to mine that arithmetic does not contain any synthetic knowledge a priori but contains only merely analytic knowledge. . . .
>
> My suggestion, which is open to correction, would therefore be to strike out section II, from pp. 54–71 and (if time does not allow undertaking the investigation required) perhaps only to bring up the importance of such an investigation, in place of the section as you had conceived it. A claim which is in such conflict with all that follows as what is contained in that

section would seem to come in useful to those who only need a pretext in order to be spared any deep investigations, indeed, even in order to maintain that there is no synthetic knowledge a priori but that the old principle of contradiction is sufficient everywhere. . . .

I hope to have the honor of discussing this with you personally.[47]

In point of fact, Schultz let himself be persuaded by Kant. The *Prüfung* itself defends the view that judgments of arithmetic are also synthetic, and that the synthetic character of arithmetical judgments rests on the axiomatic nature of arithmetic, newly recognized and illustrated. In any case, we see from this history of the *Prüfung* that near the end of the year 1788 Schultz did not yet know of the basic principles. There are two possibilities. Either Kant already knew of the axioms of arithmetic in 1788 and simply passed them on to Schultz, or else these axioms were discovered in the course of the discussion between Kant and Schultz about the nature of arithmetical judgments. Kant would really be considered the discoverer of the axioms even in this second case, since he undoubtedly was the leader in these discussions. An unambiguous proof of the first possibility would be clearer. Unfortunately, I have been unable to find any such genuine evidence, in spite of all my searching.

In the 1788 letter referred to above, Kant explicitly speaks of the view that arithmetic has no axioms, only postulates: "To be sure, arithmetic does not have axioms. . . . But it does have postulates."[48]

In any case, one could still ask whether Kant had already used the basic principles for establishing the synthetic character of the proposition $7 + 5 = 12$ without having identified them as axioms. Anyway, the theses which J. S. Beck defended for his habilitation in Halle in 1791 show that there was heated discussion about the question in Königsberg. His fifth thesis was: "It is possible to doubt whether arithmetic has axioms."[49]

For our question, two passages from the *Prüfung* itself can be cited. Schultz relates the history already known to us from the letters:

> In fact, I doubt that any empiricist could really make such a mistake as to hold that a properly demonstrated arithmetical or algebraic proposition is empirical, and, for example, persuade himself that "two times two is four" is only certain because he has not yet perceived anything different—that is, so long as he does not actually maintain that the law of contradiction itself is empirical, and is certain only insofar as one can expect, that having been proven by all past experience, it will continue to hold in the future. Rather, the opponents of our philosopher unanimously say that all propositions of arithmetic are analytic and thus that they are propositions a priori, since it would be ridiculous to appeal to experience in merely analyzing concepts. In fact, I could not be less surprised by any other objection than this one, for the idea that the propositions of arithmetic, e.g., $7 + 5 = 12$, are

merely analytic, are indeed completely identical, is so plausible that here I really admire the acute insight of our Kant for having penetrated so deeply into this apparent truth. So far as I can tell, the investigation of the real nature of arithmetic is precisely one of the most difficult problems. [50]

It would be difficult to understand such a generous reference to Kant if Schultz had wanted to take credit for discovering the axioms of arithmetic, axioms about whose significance he is clear throughout. The sentence which occurs shortly afterward can also be cited: "But first I merely want to note that I have been fortunate enough to discover the real axioms [and postulates] of time." [51] The different treatment of the axioms—the explicit reference to Kant for arithmetic, the explicit claim that the axioms of time were discovered by Schultz himself—can, indeed, be interpreted to mean that the axioms of arithmetic stem from Kant and those of time from Schultz. Basically, there can really be no doubt that these propositions are essentially traceable back to Kant, even in a formulation by Schultz.

Within the limits clearly laid down just now, I believe that I can say, then, not only that Kant ultimately pointed out the axiomatic nature of arithmetic in principle, but also that he discovered two axioms of arithmetic by actual work in mathematics, namely, the commutative and associative laws of addition. [52]

PART II

CHAPTER FOUR

Problems about Classes of Numbers

W ORK ON THE FOUNDATIONS of mathematics is not at all recent but has been continuous through the centuries. Special questions have regularly cropped up along with the general questions. Thus four problems were discussed in the seventeenth and eighteenth centuries: the negative numbers, the infinitely small, the parallel axiom, and the angle of contact (of a curve with a tangent). The really great men, no doubt, always hold themselves aloof from the questions of the day as they did from these. Thus Pierre de Fermat (1601–1665) and René Descartes (1596–1650) devoted themselves mainly to the development of analytic geometry, Isaac Newton and Gottfried Wilhelm von Leibniz to the development of the infinitesimal calculus, and Leonhard Euler (1707–1783) to the application and working out of the new methods. It shows Kant's greatness that he did not let himself also be dragged into these contemporary questions— the obvious one would have been the one about the infinitely small—but that, rather, he turned his attention to axiomatics. Except for these few great men, in fact, every mathematician of the seventeenth and eighteenth centuries wrote at least once, somewhere, sometime, about these questions of the day. Thus an unending stream of ink was spilled over the negative numbers. Now it is true, indeed, that the negative numbers did continually offer new difficulties, difficulties which lay first in discovering rules for calculating, then in the meaning of these *imaginary* quantities which had newly appeared. Thus, for example, the well-known rule *minus times minus gives plus* is easy to know but troublesome to justify. If anyone happens to have worked his way through the laws of calculation of addition and subtraction, then he can also cope easily with powers with negative bases or exponents. But then, the roots and logarithms of negative numbers present difficulties all the greater. Abraham Kästner's discussion in his *Anfangsgründe der Arithmetik* of 1758, to which Kant refers in the preface to *An Attempt to Introduce the Concept of Negative Quantities into Philosophy,*[1] or W. J. G. Karsten's two treatises in his *Mathematische Abhandlungen* of 1786, comprising 185 (!) pages,[2] are characteristic of the literature on this question.

At that time it was also believed that one could expose paradoxes in the multiplication of negative numbers. Kant and Schultz both allude to these paradoxes. In the *Anfangsgründe der reinen Mathesis,* Schultz says: "All paradox sought in the multiplication of opposed quantities is empty delusion, and that kind of talk, erroneous for more than one reason, should simply be banished from mathematics altogether."[3]

Kant's *The Concept of Negative Quantities,* which appeared in 1763, deals with negative quantities in detail, presenting his view as follows:

> A quantity is negative with respect to another insofar as it can be taken together with it only by means of their mutual opposition, namely, so that the one cancels the other insofar as they are equal. To be sure, this is really a relation of opposition, and quantities thus opposed to each other cancel out an equal in each other so that there really are quantities which can simply be called negative; rather, it must be said that $+a$ and $-a$ are the negatives of each other. However, since this meaning can always be added in thought, mathematicians have accepted the practice of calling the quantities with the $-$ in front negative quantities. Nevertheless, we must not lose sight of the fact that this label does not indicate a particular kind of thing by reference to its inner nature but only the relation of opposition it has to other things which are labelled with $+$.[4]

If this essay is examined more closely, we see that for Kant the classes of numbers, particularly the negative numbers, are to be regarded as descriptions of particular facts and not as purely mathematical developments. Schultz's *Anfangsgründe,* in particular, shows that this train of thought was followed both during and after the *Critique:*

> The theory of quantities opposed to each other is not an arbitrary fiction of the mathematician, for there really are quantities which are opposed to each other because of their nature so that they cancel each other out, partially or totally, when they are combined, e.g., wealth and debts.[5]

Formulas were then given in the discussion about the addition and subtraction of negative numbers which prepared for the commutative and associative laws of addition. The most important and, so far as I can tell, the first place is in Euler's *Vollständige Anleitung zur Algebra* of 1770 in paragraph[s 13 and] 14:

> [13. In the same way (described earlier) it is also easy to determine the value of such formulas in which both signs appear, $+$ *plus* and $-$ *minus;* as e.g.,
>
> $12 - 3 - 5 + 2 - 1$ has the same value as 5. Or one may take the sum of just the numbers with $+$ in front of them, like:

12 + 2 makes 14, and then take from it the sum of all the numbers with − in front of them, namely, 3, 5, 1, that is 9, and then 5 results, as was found before.]

14. Hence it is clear that thereby nothing at all depends on the order of the numbers as they are presented but that, instead, they can be reordered arbitrarily, so long as each one retains the sign in front of it. So instead of the formula above, $12 + 2 − 5 − 3 − 1$ or $2 − 1 − 3 − 5 + 12$ or $2 + 12 − 3 − 1 − 5$ can be written. It should be observed, however, that in the formula above the sign + has to be understood as preceding the number 12.[6]

Euler gives the general formula in paragraph 259:

If the expression $d − e − f$ be added to that of $a − b + c$, then the sum is expressed in the following form:

$$a − b + c + d − e − f$$

whereby it should be observed that nothing here depends on the order of the terms but that they can be reordered arbitrarily, so long as each one retains the sign written in front of it. Therefore, the sum above could also be written:

$$c − e + a − f + d − b.[7]$$

Remarkably enough, these formulas did not appear in the textbooks of the following two decades, the only exception to my knowledge being in the 1789 *Anfangsgründe der Mathematik* by Carl Hadaly von Hada (1743–1834):

The value of quantities with many terms remains unchanged no matter what positions their terms have relative to each other; for the value of each individual term remains unchanged, no matter how they (with their signs) are rearranged: $a + b − c = b − c + a = −c + a + b.[8]$

Kant, contrary to the views of his time, expresses himself in a very reserved way about the value of definition in mathematics in the *The Concept of Negative Quantities:*

The concept of negative quantities has long been used in mathematics and it is also of the greatest importance there. Nevertheless, the idea which most have gotten of it and the explanation they have given is astonishing and contradictory, although no inaccuracy has arisen in application, for the particular rules replaced the definition and guaranteed the use; but what may have been mistaken in the judgment about the nature of the abstract concept has remained useless and has been without consequence.[9]

Here it is clearly stated for the first time that mathematical concepts are determined by *particular rules* rather than definitions, and therefore ultimately by axioms.

There is nothing of importance in Kant's work about fractions, ratios, proportions, or progressions. The relevant chapters in Schultz's *Anfangsgründe*, on the other hand, confirm the observations already made about negative numbers. In contrast to the arbitrary fictions which always seem to be possible, a real mathematical meaning is only given to new classes of numbers when they describe particular facts.[10] There are also some Reflexions of Kant's which, however, present initial thoughts rather than ones that have been fully worked out. In Reflexion 2885 of around 1752–56, he says, "Separation is subtraction."[11]

We encounter three principal problems with the irrational and complex numbers: the representation [*Darstellung*] of irrational numbers by means of infinite expressions, the meaning of imaginary numbers, and the transcendence of certain numbers, in particular, of e and π. The first question is discussed in detail in a 1790 letter from Kant to August Wilhelm Rehberg (1757–1836):

> The $\sqrt{2}$ is expressed by the mean proportional between 1 and the given number, $= 2$. So it is also possible to think of such a number. . . . But that the mind which forms the concept of the $\sqrt{2}$ for itself at will should not be able to produce the complete concept of the number, namely, by means of its rational ratio to unity, but must let itself be satisfied with pursuing an approximation ad infinitum in determining this $\sqrt{2}$, led by another faculty, as it were. . . . It does not really seem strange to me that it must be possible to find a square root for every number, if necessary, one that is not itself a number but only a rule for approximating one as closely as required, . . . but that this concept can be constructed geometrically and is therefore not merely thinkable but can also be presented adequately in perception. . . .[12]

Kant thus sees it as astonishing that certain numbers cannot be represented finitely in arithmetic but that the same numbers can easily be represented finitely, and so adequately, in geometry. This same train of thought occurs in Reflexions 13 and 4762. Reflexions 5652 and 6434 compare the irrational numbers with the Ideas of dialectic because they are represented by an infinite series; it appears, however, that this comparison does not have any greater significance.

The seventeenth century had already decided to admit imaginary numbers into mathematics because they are indispensable to it; but they still remained alien to mathematics. They were indispensable basically because the number of roots of an equation corresponded to the degree of the equation only when the imaginary roots were included. Descartes said in his 1637 *Geometria:*

> Moreover, neither the true nor the false roots are always real but are sometimes wholly imaginary; that is, we can, of course, always imagine as many roots of an equation as I have named. But sometimes there is no quantity which corresponds to those which we imagine.[13]

Newton, in his *Arithmetica universalis* of 1707, was the first to pose the question of the number of complex roots of an equation; Colin Maclaurin (1698–1746) then pursued these researches further.[14] In this connection, Leibniz also came to speak of the imaginary roots: "So he finds an elegant and admirable escape route in that miracle of Analysis, that portent of the ideal world, practically an Amphibian between Being and Not-Being, which we call an imaginary root."[15] Kant also defends this view in letters to Rehberg and then to Schultz. Schultz himself says in the *Anfangsgründe:*

> The expressions $\sqrt{-1}, \sqrt[4]{-16}$, and in general, $\sqrt[2n]{-1}$, have as much meaning as a four-sided circle. Nevertheless, these expressions are of great use in mathematics, and so they are called impossible or imaginary numbers.[16]

It can be seen that Kant saw the difficulties of the time although he was not able to go further himself. A really general view was only achieved at the end of the eighteenth century by the geometrical representation of the complex numbers, discovered by different scholars, a representation which came to be applied generally because of Gauss's work.

The problem of the transcendental numbers arose out of the attempts to deal with the number π[17] Close approximations had been found very early. The question then arose of whether a precise representation of π was possible: first of all, whether π could be represented rationally and, if not, whether π could at least be found as a root of an equation; i.e., whether π was algebraic or transcendental according to the modern distinction. The great mathematicians of the seventeenth and eighteenth centuries certainly had no doubts about the irrationality of π, but they also assumed that π was transcendental. Remarks about this are to be found in Leibniz and Euler, but Lambert was the first to be able to prove that π is irrational. The proof depends on the fact that the representation of π cannot terminate a continued fraction but has to go on ad infinitum. Johann Schultz made this proof of Lambert's sharper in his 1803 *Sehr leichte und kurze Entwicklung einiger der wichtigsten mathematischen Theorien,* which indicates the kind of significance attributed to it in Königsberg.[18]

Lambert's proof is in the "Mémoires sur quelques propriétés remarquables des quantités transcendentes circulaires et logarithmique," published in 1768. At the end of this work, Lambert explicitly states his conjecture that neither π nor e nor numerous other quantities can ever be represented as roots of algebraic equations.[19] This work of Lambert's, however, was completely ignored by the mathematicians of his time.[20]

Anton Edler von Braunmühl (1853–1908) comments in his 1908 "Trigonometrie, Polygonometrie und Tafeln":

> The whole character of Lambert's reasoning, with his intention of being absolutely precise, differed so completely from the activity of his contemporaries, which was directed almost exclusively to the formal expansion of mathematics, that it can well be understood why no one else paid attention to it.[21]

In particular, Wolff did not accept the existence of the transcendental numbers since he thought that all irrational numbers were algebraic: "Irrational numbers are as one straight line to another straight line."[22]

Kant followed the work of Lambert much more attentively. That can be gathered not only from the work of Schultz already cited but also from the explicit reference to Lambert's proof in the *Critique of Pure Reason,* even if it is only cited as an illustrative example:

> Has it ever been suggested that, because of our necessary ignorance of the conditions, it must remain uncertain what exact relation, in rational or irrational numbers, a diameter bears to a circle? Since no adequate solution in terms of rational numbers is possible, and no solution in terms of irrational numbers has yet been discovered, it was concluded that at least the impossibility of a solution can be known with certainty, and Lambert has given the required proof of this impossibility.[23]

The mathematicians of the day passed over Lambert's proof, and it was only taken up again by Gauss. Nevertheless, Kant recognized the significance of Lambert's work, so I do not think it can be said that Kant had no interest in mathematics or that he did not understand anything about mathematics. It is particularly worth noting that Wolff completely passes over the question of transcendentality, so that in no way did Kant acquire his knowledge from Wolff. Moreover, it would be completely unintelligible if Kant had not read all the works of Lambert, who was so close to him. The passage just cited from the *Critique* shows, however, that Kant appreciated the fundamental discovery of Lambert with sure insight.

CHAPTER FIVE

Combinatorics and the Idea of a Systematic Ontology

T HE MATHEMATICAL DISCIPLINE which investigates the different possible combinations of a given number of elements, with or without regard to their order, is called combinatorics. The question of how many different arrangements there are for twelve people to sit around a table was a problem even in antiquity.[1] Establishing the formulas required kept the mathematicians of the seventeenth century very busy. Pascal, Leibniz, and Jacob Bernoulli (1654–1705) all wrote special works about it.[2] Bernoulli's *Ars conjectandi* was in Kant's library. It was believed that the foundations of all mathematics had been found in combinatorics, and a special school of the discipline developed, mainly in Germany, under the leadership of Karl Friedrich Hindenburg (1741–1808).[3] The methods of combinatorics did not really fulfill these high expectations. Nevertheless, in the eighteenth century, up to the beginning of the nineteenth, combinatorics played an unusually large role in the textbooks in the discussions about foundations.[4] Combinatorics then acquired a fourfold importance for logic and ontology. First, the complex concepts were viewed as combinations of simple concepts; second, the judgment was viewed as a combination of two concepts, namely, of the subject and the predicate; third, the syllogism was regarded as a combination of judgments. The fourth problem went beyond these, whether it was possible to establish a system of signs for a particular field of knowledge so that all the relevant possibilities of the field considered could be discovered by forming all the possible combinations of the signs. The only one of these four problems still a live issue is the least important one, namely, to find the possible number of methods of inference by combining judgments. In this chapter I will try to show the significance of these problems for Kant, and thereby to establish connections between Kant and Leibniz.

To begin with, what is basic is the interpretation of complex concepts as combinations of simple concepts. We really already have such an approach when a species is conceived of as a joining of the genus proximum and the differentia specifica. Ramón Lull (Raymundus Lullus, ca. 1232–1316) undertook the first explicit attempt of this sort.[5] The work most important for us is the youthful writing of Leibniz which appeared in 1666, *De Arte combinatoria,* in which the mathematical and philosophical ideas are all mixed together. The ontological propositions which interest us appear as applications of the combinatorial formulas given by Leibniz. The reflections there about the way to determine the number of possible forms of the syllogism are relatively uninteresting. The idea, following Lull, about the determination of the number of all propositions and, in particular, of all true propositions, carries us much further:

> A proposition is composed of subject and predicate, and so all propositions are combinations. It is then the business of inventive logic,[6] so far as it concerns propositions, to solve this problem: 1) Given a subject, to find its predicates. 2) Given a predicate, to find its subject. In each case, this is to cover both affirmative and negative propositions.[7]

The combinatorial interpretation of complex concepts, however, proves to be the most important. If there are such complexes (*homo* equals *rational animal*), then, according to Leibniz, there must also be simple concepts, but further, all complex concepts must be conceivable as combinations of the simple ones:

> Let any given term be analyzed into formal parts, i.e., let its definition be given and let these parts again be analyzed into parts, i.e., let there be a definition of the terms of the definition, down to simple parts, i.e., indefinable terms.[8]

After Leibniz has thus split up all concepts into basic or elementary concepts as elements, he constructs all possible combinations in twos, in threes, . . . and so constructs all complex concepts.

Everything that, according to Leibniz, is associated with *scientia generalis* (universal science) or *ars characteristica* (art of signs) is only an elaboration of this basic idea which he tries to work out with algebraic, arithmetic, geometric, and mechanical methods. For example, take arithmetic. If the elementary concepts are represented by the prime numbers, complex concepts are then represented by products of the prime numbers.[9]

Leibniz's analytic theory of judgment is built upon this. If any judgment "*S* is *P*" is interpreted as a combination of two concepts, then the concept of the subject is also interpreted as a combination of *elementary* concepts. But then the judgment consists in the fact that the predicate extracts a single component from this complex concept of the subject; but in turn this means that the judgment assumes the form "(*AB*) is *B*." In this way, any judgment can and must be necessary because the component concept *B* must necessarily be predicable of the complex concept (*AB*), as say, *rational* of *rational animal*. Thus Leibniz arrives at the final statement: The predicate is in the subject.

> Generally every true proposition . . . can be proven a priori with the help of axioms or of propositions which are true in themselves and with the help of definitions or ideas. For whenever the predicate is truly asserted of the subject, there is certainly thought to be some real connection between the predicate and the subject, just as in any proposition: *A* is *B* . . . *B* certainly should exist in *A,* or the concept of it should be contained in some way in the concept of *A*.
> So the predicate or that which follows always exists in the subject or that which precedes, and the nature of truth universally consists in this very fact.[10]

Here we are only dealing with mathematical judgments and and can confine ourselves to considering judgments whose subject-concepts are finitely complex. Subject-concepts which are infinitely complex are, for Leibniz, the defining characteristic of contingent judgments, which can be made a priori only by a divine mind in a position to analyze infinitely great complexity. This theory of Leibniz's was first made clear by Couturat, who investigated this broad problem in all its ramifications, for its mathematical, logical, and ontological bearings.[11]

If we now compare this approach of Leibniz's with what Kant contended was analytic judgment, we see that they are wholly identical. Thus it is untenable to criticize Kant's theory of judgment for not being precise about the way a predicate-concept can be contained in the subject-concept—in fact for never having made precise what this *being contained* is supposed to mean. Rather, in his account of analytic judgment Kant was entirely accurate in his presentation of Leibniz's theory, and we can refer back to Leibniz for details of the manner in which a predicate-concept is contained in the subject-concept.

This approach of Leibniz makes possible an all-embracing relation between an area of knowledge and a system of signs. If there is such an area in which finitely or denumerably many basic concepts exist, and if all complex concepts in this area of knowledge are combinations of these basic concepts—it remains to be seen whether only a single area or the whole of knowledge exhibits this structure—then such an area can be represented by designating the concepts by means of basic symbols [*Grundzeichen*]. Additional development can be confined to combining these basic symbols, and *every* such combination must designate a complex concept. The difficulties, therefore, arise in two ways: To begin with, the basic concepts must exist and be ascertained; but then all the perplexities lying in the relation between symbol and symbolized enter later on.

Incidentally, if the analytic judgment of Kant has reference to Leibniz's theory of judgment, then we can test what is often affirmed, namely, that Kant had not really studied Leibniz himself but had become acquainted with him only through Wolff. If such a view is prima facie implausible, given Kant's practically unlimited capacity for work, it proves to be unsatisfactory with respect to the present problem. Wolff did not hold Leibniz's analytic theory of judgment. In fact, Wolff said that every proposition [*Aussage*] was somehow contained in something else: "Things which always exist in something can be described absolutely and conversely."[12]

The Leibnizian approach, however, is basically shifted by Wolff by his reverting to the scholastics; in the description of a thing he distinguishes between essentials, attributes, and modes. Judgment is, moreover, identical for Wolff only in the rare case in which a thing is defined by its essentials and the predicate of the proposition is formed by taking a single essential determination from the totality of essentials which form the subject-concept: "If the subject is defined by essentials, and one or more of them are predicated of it, the proposition is an identical one."[13]

Such a judgment would, therefore, conform to Leibniz's description. According to Wolff, one may define a subject in another way, say by a sufficient number of attributes, as when he supposes "If the subject is defined by *attributes*"[14] In such a case, we can, so to speak, make inferences in both directions, i.e., from the determination of what is essential down to the attributes or up from the attributes to what is essential. Wolff does not want to call this identical any more. In order to make the distinction clear, we will consider what is so often the case in mathematics, namely, that a concept is defined in two different ways. Let the first definition be (a, b, c), the second (f, g, h). Then the judgment that (a, b, c) is a or the judgment that (f, g, h) is f is an identical one for Wolff as well as for Leibniz. But, on the contrary, it is different for the judgment that (a, b, c) is f or the judgment that (f, g, h) is a. Leibniz now introduces the following considerations about these judgments. Since both definitions (f, g, h) and (a, b, c) define the same object, the concepts used for the definitions must also be complex concepts which in turn must be decomposed into the same basic concepts on further analysis. Let $a = kl$, $b = mn$, $c = op$, and $f = kn$, $g = mp$, $h = ol$; then the two definitions are resolved into $(a, b, c) = (kl, mn, op)$ and $(f, g, h) = (kn, mp, ol)$. The four judgments are represented in this fashion:

(a, b, c) is a as (kl, mn, op) is kl

(a, b, c) is f as (kl, mn, op) is kn

(f, g, h) is f as (kn, mp, ol) is kn

(f, g, h) is a as (kn, mn, ol) is kl

Thus for Leibniz all judgments are really judgments of identity. Such a resolution of two different definitions into the same basic concepts is, however, quite impossible for Wolff. He has, indeed, entirely banned such a theory of judgment by the distinction between essentials, attributes, and modes, for Leibniz's theory of judgment requires, on the contrary, concepts which are wholly alike in form. Only the distinction between simple and complex concepts is possible in it. So Kant could hardly have learned about this theory of judgment from Wolff.

Herman Schmalenbach (1885–1950) in his 1921 *Leibniz* is probably the first who broke with the usual view that Kant saw Leibniz only through Wolff, pointing out, first of all, that the young Kant knew Leibniz extremely well, being an advocate of his, and that Kant's own work must certainly be viewed as a debate with that philosopher.[15] He further pointed out that Kant had a far better understanding of Leibniz on important basic questions, such as that of the preestablished harmony, than had Wolff himself.[16] Our account of analytic judgment comes, as we saw, to the same conclusion on this particular point.

Kant held Lambert in unusually high esteem; indeed, Lambert was, for him, the only contemporary he found to agree with his own philosophical aims. This great regard is sufficiently indicated in Kant's letters to Lambert. In addition, in 1766 Lambert wrote to Georg Jonathan Holland at a time when he himself was already famous and Kant was still an unknown privatdozent: "Shortly after I sent off my most recent letter, I received a short treatise, *Dreams of a Spiritseer.* This philosopher, whose way of thinking is closest of all to mine. . . ."[17]

Kant wanted to dedicate the *Critique of Pure Reason* to Lambert in order to acknowledge this relation. Around 1776 he wrote in a draft of a letter to Lambert: "You have honored me with your letters. The effort to form a concept of the method of pure philosophy has caused me a series of reflections."[18]

If the question is asked in what ways Kant and Lambert were in agreement, it is not easy to answer. First, some more statements of Kant's can be cited. In Reflexion 4966 of around 1776–78, he says, "Lambert analyzed reason, but there was still no critique."[19]

Reflexions 1629, 4893, 4900 express similar ideas.[20] Kant stated explicitly what he meant by such an analysis in the introduction to the Analytic in the *Critique of Pure Reason:*

> By the analytic of concepts I do not mean their analysis or the usual procedure in philosophical inquiries, of analyzing the content of any given

concept . . . , but the analysis of the *faculty of intellect itself,* which is seldom attempted. [21]

Accordingly, the whole of the next paragraph refers mainly to Lambert:

When one puts a faculty of knowledge into play, then different concepts manifest themselves according to the differing circumstances, and make this faculty knowable, and let themselves be collected more or less in full, according to whether the observations of them have been made over a longer time or with greater *acuteness*. But when the inquiry is made in this almost mechanical way, one can never be sure when it is completed. Furthermore, the concepts thus discovered by chance show no order or systematic unity but, in the end, are merely arranged in pairs according to similarities, and in series according to the sizes of their contents, from the simple to the more complex—an arrangement which is anything but systematic, although formed according to a method in a certain sense. [22]

Thus, Kant's criticism that Lambert *only analyzes* means that he acknowledges the task of philosophy is to search for the basic concepts, but he rejects Lambert's method of analyzing the common concepts presented to him and so hoping to derive the basic concepts by methodical observation, as it were. We have to search for basic concepts, says Kant, not as concepts happen to occur but in their place of origin, if we want to derive them all.

Let us make a short survey of Lambert in order to test the validity and significance of Kant's criticism. The main philosophical works of Lambert are the 1764 *Neues Organon, oder Gedanken über die Erforschung und Bezeichnung des Wahren und dessen Unterscheidung von Irrtum und Schein* (The New Organon, or Thoughts on the Search for and Description of the True and Its Distinction from Error and Appearance) and the 1771 *Anlage der Architectonik, oder Theorie des Einfachen und Ersten in der philosophischen und mathematischen Erkenntnis* (The Construction of Architectonic, or Theory of the Elementary and the Primary in Philosophical and Mathematical Knowledge). [23] The numerous works by Lambert about logistic are in the *Architektonik* and in journal articles.

To begin with, Lambert takes over from Leibniz the basic distinction between simple and complex concepts:

The next conclusion to be drawn is that if we want to be sure that a concept contains nothing contradictory and hence is a real and possible concept, we must be able to show that it is composed of simple concepts in some permissible way. [24]

Thus, a basic science arises in philosophy whose task it is to search for simple concepts and whose method consists in an examination of human concepts:

> Our scientific knowledge fully and in the strictest sense . . . would be a priori if we knew the basic concepts collectively and had expressed them in words and knew the basic principles of their combination."[25]

> If we want to search for every individual simple concept, all the human concepts must be examined together.[26]

Lambert's aim, therefore, is to search for the basic concepts in their entirety. He explicitly leaves it open whether this is possible:

> Since the realm of truths, just like the realm of possibilities, extends to infinity, if this is the goal, human knowledge will always remain incomplete. So, for example, we can search for and count the simple concepts on which our whole knowledge is based. But we could miss just as many, as the blind the concepts of the colors."[27]

In the *Architektonik* Lambert argues against Locke and Wolff.[28] Locke had, to be sure, found the simple concepts but had failed to draw the possible conclusions, while Wolff had developed the method to greater subtlety but had not noticed the distinction between simple and complex concepts. (I have already pointed out that this basic distinction of Leibniz's was no longer made by Wolff.) Lambert then offers the following list of basic concepts in the *Neues Organon:* "Consciousness, Existence, Unity, Duration, Succession, Volition, Extension, Motion, Force."[29] He calls the procedure for deriving the simple basic concepts from the concepts as they are presented to us *analysis.* When the basic concepts are derived, the synthesis begins—the construction of the complex concepts from the simple ones. This proceeds by going through all combinations of basic concepts and eliminating those combinations having contradictions. Two goals are achieved by such a procedure: all true concepts are derived in their totality, and in addition the method acquires the certainty and clarity of mathematics.

> It is easy to see that with this a combination is possible by which all changes made from a particular number of concepts can be determined. And if this is done, it is indisputable that not only can everyone be satisfied but also that all the gaps that remain can be filled.[30]

> If we begin with simple concepts, we can achieve absolute certainty by avoiding leaps in our uniting or combining, as far as we go. For simple concepts can be thought of in themselves . . . , and error can only enter in their uniting or combining. . . . Accordingly, if we make sure of the

possibility of combining or uniting for each single part we combine or unite, we proceed step by step; and since, accordingly, any gap is avoided, the possibility of the representation is guaranteed, as far as we go, and the consciousness of each one of these steps makes the certainty absolute.[31]

We take this system or realm of truth here in such a way that if we knew how to represent all truths in our way, they would form the realm of truths. Accordingly, we regard the entire system of all concepts, propositions, and relations which are possible, with all its interconnections from the very first, and we look at what we already know in some way as parts and individual pieces of this system, because we can have the outline of the entire structure before our eyes and can thus test every individual piece in this way, as often as we find new pieces and want to relate them to what we have already found.[32]

After this brief survey, we can now judge how much Kant and Lambert agree with each other and where the difference between them begins. The basic goal is common to both of them, namely, to find the system or realm of truth, as Lambert puts it, or the system of pure reason, as Kant does, by first establishing the system of basic concepts and then deriving the complex concepts in their entirety from these basic concepts. Kant goes beyond Lambert's approach in three points: he chooses another method; he is convinced the system of pure reason is complete; he divides the homogeneity of a priori knowledge (still firmly retained by Lambert from Leibniz) in that he thinks there are two sources of knowledge. He has made the first point clearly enough in the passage quoted earlier, and does not want to derive the system of basic concepts by examining in detail the concepts used until now collectively; rather, he looks for the totality of human concepts in their place of origin. Secondly, Kant maintains that the system of pure reason is absolutely complete. I will cite later some of Kant's observations about this. We will be concerned in the following with the third point, the consequences of the division of a priori knowledge into two branches.

Two questions arise. We have just seen that the method of philosophy for Lambert coincides, in the final analysis, with the method of mathematics, a danger which threatened not only Lambert but others as well. It is doubtful whether Kant, who followed these positions of Leibniz and Lambert to an astonishingly great extent, escaped the danger of mathematizing philosophy. In itself, this question already appears strange in view of the passionate fight Kant waged against the merging of mathematics and philosophy. In what follows, however, we will meet with some most remarkable passages which require dealing with the question.

The second question is on opposite lines. Undoubtedly, by changing his way of putting it Kant was convinced that he had succeeded in ensuring the completeness of his basic concepts and thus, at the same time, in laying the foundations of a system of pure reason. But now one must ask whether the heterogeneous character of what he recognized as pure knowledge, with its origin in two different sources, does not undermine this laboriously achieved completeness.

The *ars characteristica universalis* (general art of signs) projected by Leibniz was much discussed in the eighteenth century; at the very least, Kant was acquainted with it through Lambert. At the beginning of his philosophical work in his 1764 *Enquiry into the Clarity of the Principles of Natural Theology and Morals,* Kant brusquely rejected such an approach:

> Here (in philosophy) neither figures nor visible signs can express thoughts or their relations; moreover, it is impossible to replace abstract reflection by using the method of rearranging signs according to rules so as to substitute the idea of the things themselves by the clearer and easier one of signs. The universal must, instead, be considered *in abstracto*.[33]

Because of this realistic rejection, Kant pours biting sarcasm on the logicist efforts of Leibniz in his 1755 *A New Exposition of the First Principle of Metaphysical Knowledge:*

> I confess that what this great philosopher said reminds me of the will of that old man in Aesop's fables who, when he was just about to give up the ghost, disclosed to his children that he had hidden the treasure somewhere in a field [but suddenly expired before he had indicated the location. This led his sons to turn up the soil industriously and cultivate it by digging until, although their hopes were frustrated, they were certainly made richer by the fertility of the land. I venture to say that this is the only fruit to be expected from the investigation of Leibniz's famous device, if, that is to say, there are any who are prepared to persevere in the task.] But if open confession is right (as it surely is), I am afraid that what the acute Boerhaave suspects to be the case in chemistry with regard to the best alchemists [has likewise befallen that distinguished man].[34]

It is certainly harsh when Kant compares Leibniz to the alchemists in this passage. The same attack is repeated in the last of the minor early writings, *The Basic Reason for the Distinction of Regions of Space* of 1768:

> The renowned Leibniz had many genuine insights with which he enriched knowledge, but the world waited in vain for him to execute even greater projects. Whether the reason was that his efforts still appeared incomplete to him, a delicacy characteristic of distinguished men which has deprived learning of many valuable tidbits time and time again, or whether what happened to him is what Boerhaave speculated about great chemists, that they often claimed they had performed feats as if they actually had, whereas

they really only had the conviction and confidence that they could do something once they had decided to do so and that they could not fail—I do not wish to decide that here.[35]

This abrupt rejection of an *ars characteristica universalis* is, in fact, nothing else than the rejection of mathematical methods in philosophy. It is thus absolutely necessary for Kant, who had already distinguished sharply between mathematics and philosophy in his early writings. Now, since this rejection of mathematical methods in philosophy continued to be maintained in the well-known chapter on "The Discipline of Pure Reason" in the *Critique of Pure Reason*,[36] it should be impossible that Kant would have given up his opposition to Leibniz's *ars characteristica*. And yet he went so far as to use, or at least try to use, these purely mathematical methods of the *ars characteristica universalis* in his system. This surprising change of Kant's mind can be inferred, first of all, from the Reflexions and the correspondence. From there on, it will no longer be possible to explain away numerous passages in his works, in particular, in the *Critique of Pure Reason* and the *Metaphysical Foundations of Natural Science*, but they will have to be taken as they actually occur.

The scornful rejection is still to be found, if in somewhat veiled form, in the *Critique*, that is to say, in the chapter about the difference between mathematics and philosophy:

The great success which reason has by means of mathematics quite naturally arouses the expectation that it, or at least its method, will also be successful in other fields besides that of quantity; for it is able to realize all its concepts in perceptions which it can provide a priori and by which it becomes, so to speak, master of nature. . . . Nor do the masters of this art seem to lack any confidence in this procedure—nor do ordinary mortals in their great expectations from their skill—should they apply themselves to the project.[37]

However, the following three Reflexions sound quite different.
Reflexion 4937, from around 1776–78:

It is of the greatest importance to make rational science technical. The logicians have tried in vain to do this, with their syllogistic as a factory. The inventors of the algorithm have only been successful with respect to quantity. Should it not also be so in the critique of pure reason, not in increasing knowledge but in clarifying it? By labeling, every concept can be given its function through the technical method, or rather, the functions themselves can be expressed in themselves and in contrast to each other. (Algebra expresses them only in contrast to each other, perhaps also so in the transcendental algorithm. Errors and oversights can only be prevented by it.)[38]

Reflexion 4938, from around 1776–78:

Mathematicians have believed that they have achieved something superior when they work with objects of pure reason. But it is to be regretted that they are doing a thankless task when they take this knowledge as objective. But they can devise something very useful by working on the critique of reason. Holland.[39]

Reflexion 5047, from around 1776–78:

Mathematics may well be useless in the discovery of objective, philosophical propositions because it cannot judge how certain the data are. But once this is settled, a mathematical brain can discover a transcendental analysis just as it can a universal arithmetic.[40]

All three Reflexions have the same meaning. Mathematics should be used in philosophy as an aid. The expressions are amazing: "transcendental analysis" and "transcendental algorithm." Since an algorithm is a procedure for calculation, Kant is demanding a transcendental calculating procedure in philosophy. These Reflexions point to mathematics as an aid, indeed almost as a method of philosophy, and signify Kant's extraordinary inclination toward mathematics, beyond his original rejection.

For further evidence, I first draw upon Kant's correspondence. Letters to Herz, Schultz, Beck, Karl Leonhard Reinhold (1758–1823), and Hindenburg have to be considered. They stretch over a period of more than twenty years and thus cannot have arisen from a mood he was in but once.

Kant to Marcus Herz, 1773:

I don't believe that many people have tried both to sketch out an entirely new science according to its leading ideas and at the same time work it out fully. . . . But I have bright hopes about this notwithstanding, . . . namely, to give philosophy a direction permanently much more advantageous than that of religion and morals; but also at the same time thereby to give it a form which can lure the reserved mathematicians into thinking that its cultivation is possible and valuable.[41]

Thus, just at a time when the first outline of the *Critique* had taken concrete form, Kant hoped to make the introduction of mathematical methods into philosophy possible. After the *Critique* was finished, these hopes continually became stronger. Kant wrote thus to Schultz in 1783:

These, and the other properties mentioned in part, of the table of concepts of the intellect still appear to me to contain material for a discovery which might be important . . . which is reserved for a mathematical mind like

yours: to put an *ars characteristica combinatoria* into practice. . . .
Perhaps with your acuteness, with the aid of mathematics you will find a
bright prospect here where I see only something as if veiled in a mist.[42]

Schultz answered immediately:

The clever [*sinnreiche*] idea which you, honored sir, were pleased to give
me with the aim of applying the table of categories for the discovery of the
ars characteristica combinatoria is just wonderful. But I don't know of
any man besides you whose creative genius would be adequate for the
execution of such a plan.[43]

I select only one more from among the other letters which say the same thing,
one to Beck.
Kant to Beck in 1791:

I still haven't completely lost hope that mathematics can lead to new
approaches to the critique and survey [*Ausmessung*] of pure reason, even if
this study can't shed any new light on mathematics itself. It might do this
by considering its methods and heuristic principles together with the
requirements and desiderata associated with them. And perhaps mathematics might be able to contribute to this critique and survey of pure reason
by giving new ways of presenting its abstract concepts and even by creating
something similar to Leibniz's *ars universalis characteristica combinatoria*. For the table of categories and that of the ideas, among which
the cosmological somethings appear similar to the impossible roots,[*] are
indeed enumerated and so defined by concepts with respect to all possible
uses of reason as only mathematics can require in order to see how much
they could increase our knowledge or at least introduce clarity into it.
    [*If I try to find the unconditioned and supreme basis of the whole series
of appearances according to the principle "everything in the series of
appearances is conditioned," it is as if I were trying to find the $\sqrt{-2}$.][44]

Because of these passages, we can no longer overlook the fact that Kant gave
up his originally strong opposition to Leibniz on logistic and that, on the
contrary, he himself saw in this method the way to completing his own system.
We will, accordingly, deal briefly in the next section with Kant's relation to
logistic in order to be able then to consider the position and significance of
logistic in the system of pure reason.

The approach with which Leibniz had begun logistic was enthusiastically
carried on in the eighteenth century. Lambert worked on this problem particularly hard. Almost all the works of the other authors are cited by him. At the
very least, Kant must have known about Lambert's papers. In spite of this,
scarcely any inquiry into the problem of logistic is to be found in Kant. *An*

Attempt to Introduce the Concept of Negative Quantities into Philosophy (1763) might in a way be included here. Some *Reflexions* are also to be found associated with this, for instance, Reflexion 1946: "Beautiful $+$; Not-Beautiful . . . 0; Ugly $-$."[45] Or Reflexion 3711: "Therefore Nothing is the negative." [*Nihil negativum*.]; "$P - P = C \times P$."[46] But it does appear as if these minor attempts at logistic had no particular significance for Kant. Nor is there any kind of attempt at logistic in the work of his mathematically inclined friends and students, Johann Schultz, in particular. We have already seen earlier how decisively he rejected Kant's suggestion that he sketch out a combinatorial logistic.

Throughout his life Kant envisioned a system to which he gave different names, such as "system of pure reason," "systematic metaphysics," "systematic ontology." Here I want to raise the question of what Kant envisioned as the content and method of such a system of pure reason and whether it can be determined why it was not worked out. In this section I will first put together Kant's statements about completing the system, then state its outlines, and finally sketch out one particular part, the systematic ontology. In the next section I then examine the difficulties standing in the way of completing this system with respect to this particular part.

From the beginning of his work, Kant had as his goal a system which would be the final one and thus bring to a close the endless quarrels of philosophers. The proud words of section 7 of the preface to the *Thoughts on the True Evaluation of Living Forces* of 1747 can only be interpreted in this way: "I have already sketched out the path I want to follow. I will set out on my journey and nothing will prevent me from continuing it."[47] When Kant wrote these words at the age of twenty-three, he could only have been thinking of a definitive solution. The announcement in *The Concept of Negative Quantities* of 1763 refers to possessing particular knowledge which has already been acquired:

> I have reflected on the nature of our knowledge with respect to our judgments from grounds and consequences and I will set out the results of these observations some day. . . . Until then, those whose pretensions to insight are boundless will try to use the methods of their philosophy to see how far they get in questions of this sort.[48]

The explicit announcement is found for the first time in the 1766 *Dreams of a Spiritseer:*

> And the philosophers will together inhabit a common world, the same one which the mathematicians have already occupied for a long time, and this important event cannot be delayed much longer, insofar as we can trust

certain signs and portents which have been appearing over the horizon of knowledge for some time.[49]

This passage leads directly into the difficulties which Kant never completely overcame and which perhaps cannot be overcome; it says very bluntly, in fact, that philosophy should be raised to the level of a universally valid science, following the example of mathematics. Is there then no danger that the basic Kantian distinction between mathematics and philosophy again becomes blurred? It is not entirely clear what the "signs and portents" are supposed to refer to. Since the work criticizes Wolff and Crusius with unusual sharpness, these two do not come into question. If Kant's own preliminary work is not to be understood as these "signs and portents"—which is difficult to believe—then only Lambert could be the man referred to here. With respect to the content, moreover, the realm of truth that Lambert strove for could very well be meant. Now Kant does present the most important and basic parts of such a definitive system in the *Critique of Pure Reason,* in particular, the complete table of pure basic concepts: "This then is the list of all the pure concepts of synthesis which the intellect has in itself a priori."[50] These are also identified as root concepts [*Stammbegriffe*], elementary concepts [*Elementarbegriffe*], basic concepts [*Grundbegriffe*].[51]

Thus Kant has now taken the definitive step beyond Lambert; doubt is laid aside, the system of pure concepts is complete. Moreover, after the Kantian turn, how could reason be blind *to* itself when it finds the pure basic concepts *in* itself? Perhaps the explanation suffices that Kant was thoroughly persuaded of its completeness and definitiveness; the decisive evidence is indeed known. The *Critique of Pure Reason,* however, does not fully present the system which is complete in itself.

> With regard to these it should be remarked that the categories, as the true *root concepts* [*Stammbegriffe*] of pure intellect, also likewise have their pure *derivative concepts* which could by no means be passed over in a complete system of transcendental philosophy, but which in a merely critical essay I can be content with merely mentioning.[52]

Kant speaks briefly of a "system of pure reason." Just before this, he promises to present a complete system: "Since I am not concerned here with the completeness of the system but only with the principles for a system, I save this supplementary work for another occasion."[53]

The preface to the first edition also promises this system: "I myself hope to produce such a system of pure (speculative) reason with the title of *Metaphysics of Nature*. It should not be half as long yet incomparably richer in content than this present critique."54

This idea is also found in numerous other passages in the *Critique of Pure Reason*. It may suffice here to cite the concluding statement:

> Only the *critical* way is still open. If the reader has been kind and patient enough to accompany me along this path, he may now judge for himself whether he cares to help in turning this path into a highway and so whether it may not be possible to achieve before the end of the present century what many centuries have not been able to accomplish: namely, to completely satisfy human reason in its thirst for knowledge about matters which have been its concern in every age, though hitherto in vain.55

These words, written about 1780, are proud ones. The questions of reason, futile until then, are supposed to be completely answered even before the end of the century within twenty years. And Kant goes even further; he believes that he has accomplished so much of the preliminary work that the individual reader can take an active role in completing the system. The similarity to mathematics appears here once more: The development of a field of knowledge always occurs with the collaboration of everyone after the founding of it by a creative genius.

The *Metaphysical Foundations of Natural Science* of 1786 is part of the work for the creation of such a system, as Kant indicates explicitly.56 In addition, reference to the work still to be done is continually found in his letters.

Kant to Moses Mendelssohn (1729–1825) in 1783:

> But before this I'm thinking of gradually working out a textbook of metaphysics according to the basic critical principles above, with all the conciseness of a handbook for the benefit of academic courses, and of completing it at some indefinite time, perhaps in the fairly distant future.57

Kant to Johann Bering (1748–1825) in 1786:

> If I succeed in this enterprise [the second edition of the *Critique*] as I'm now sketching it out, it will be within the capacity of almost every thoughtful person to sketch out a system of metaphysics according to it. So I'm going to postpone my own work on it until some other time in order to have the time now for the system of practical philosophy.58

Kant in *Some Remarks on Ludwig Heinrich Jakob's Examination of Mendelssohn's Morgenstunde* of 1786, an introductory essay to Jakob's work, concludes:

Fig. 1. Outline of the System of Pure Reason

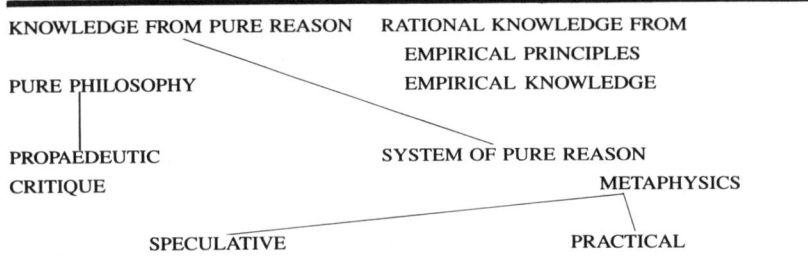

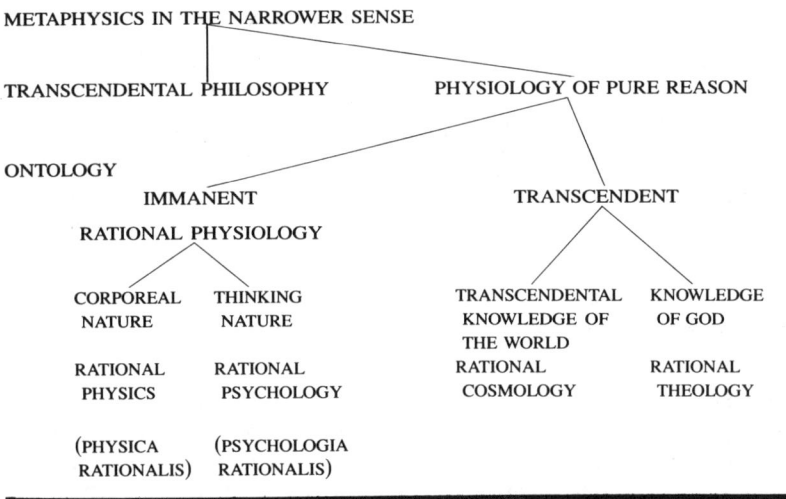

The way things in metaphysics stand now, the evidence for adjudicating its quarrels has by now made us nearly ready for a verdict so that we only need a little patience and nonpartisanship in judgment to live to see them straightened out once and for all.[59]

Kant wrote to Herz in 1787:

I've let myself in for a business in my philosophical work which at my age is pretty difficult and far-reaching; but I'm making such good progress with it, particularly with what deals with what's left, which is what I am working on now, and I have such good hopes of bringing things in metaphysics onto a secure track, that it encourages me and gives me strength to complete my plan.[60]

And he wrote to Beck in 1792:

> Besides, I've already made a sketch for getting around this difficulty with a
> system of metaphysics and to begin with the categories in their order
> (whereas earlier I've only described the pure perceptions of time and space
> in which objects are given, without investigating how they are possible).
> After finishing the exposition of each category, e.g., of quantity and all the
> predicables contained in it together with examples of its use, I then prove
> that experience of objects of the senses is possible only insofar as I assume
> a priori that they all have to be thought of as quantities; and so with all those
> remaining.[61]

These letters are sufficient to show that neither the *Critique of Pure Reason*
nor the *Metaphysical Principles of Natural Science* present the system fully. But
they show well that the further development in the *Critique of Practical Reason*
and the *Critique of Judgment* does not present the system planned. Undoubtedly, the *Opus Postumum* has the drift of this system; and Kant began it shortly
after 1790. How important this work is for our question cannot be decided so
long as there is no really dependable edition available. Moreover, this work
remained unfinished.

In one particular chapter of the *Critique,* the Architectonic of Pure Reason
(A840ff. = B868ff.), Kant gives an outline of the system of pure reason (see
fig. 1).

The *Advances in Metaphysics* [*Fortschritte der Metaphysik*] has the following divisions (according to the pagination of the K. Vorländer edition in the
Philosophische Bibliothek):[62]

System of pure theoretical philosophy

Metaphysics of nature	p. 122	(292)
Metaphysics of morals	p. 85	(260–61)
Mathematics	p. 85	(260–1)

1. Stage

Theory of knowledge	p. 99	(273)
Sure progress	pp. 99, 118	(273, 289–90)
Theoretical, dogmatic	pp. 99, 108	(173, 280–81)
Dogmatism	p. 89	(264)
Propaedeutic	p. 84	(260)
Critique of pure reason		
Transcendental philosophy	p. 84	(260)
Ontology	pp. 84, 108	(260, 280–81)

Rational physics   Rational psychology

Limits of Experience

2. Stage

Theory of doubt	p. 99	(273)

Standstill	pp. 99, 108	(273)
Sceptical	pp. 99, 108	(273, 280–81)
Scepticism	p. 89	(264)
Cosmology	p. 108	(280–81)
physica rationalis		psychologia rationalis
3. Stage		
Philosophy	p.99	(273)
Going on	p. 99	(273)
Practical, dogmatic	pp. 99, 108	(273, 280–81)
Criticism	p. 89	(264)
Completion	p. 108	(280–81)
Theology		
a. Theology	pp. 109, 133–38	(281, 301–6)
God	p. 126	(295–96)
b. Moral theology	pp. 138–40	(306–8)
Freedom	p. 126	(295)
c. Psychology	pp. 140–42	(309–10)
Immortality	p. 126	(295–96)

Although Kant continually emphasizes that the system of pure reason is complete and definitive, the worst defect of the system is that it has not really been presented in this definitive form. It would certainly be easy to apply to Kant himself the sarcastic comparison to alchemists with their unfulfillable promises which Kant knew how to apply to Leibniz so well. It is, however, more fruitful to look for the reasons why Kant did not finish the system. First, one should go into the difficulties which arise, say, from the mutual relationship of ontology, physiology, and theology. Then one could point out that the two outlines of the system given above do not coincide as would be necessary. It would also be possible to stick to the ontology and tackle the table of categories, either by substituting other categories for the individual ones while keeping the number at twelve or by adding or subtracting basic concepts. If it is held as a fundamental that a system is possible, then one can follow the approach of the German idealists. We can seek the original and basic concepts for the Kantian system from which everything else can be derived, as, for example, Johann Gottlieb Fichte (1762–1814) did; we can also try other methods of construction and, like Hegel, reject the inflexible scheme of the Kantian concepts and replace it with dialectical movement. What interests those of us, however, who have doubts that a system of pure reason is possible is the question whether Kant's indisputable desire for a system does not contradict his most basic principles so much that the system had to remain incomplete. We therefore ask whether Kant was unable to complete the system because of his age or inability to do so, or was it intrinsically impossible to complete it according to Kantian principles? When I try to investigate the aporia in Kant's system in what follows, I know that this initial attempt will not achieve the clarity which is both possible and necessary.

After all that Kant has presented, what is really still lacking in the system as a whole? As earlier noted, Kant himself identified what was missing, explicitly and precisely, in the *Critique of Pure Reason:*

> With regard to these it should be remarked that the categories, as the true *root concepts [Stammbegriffe]* of pure intellect, also likewise have their pure *derivative concepts* which could by no means be passed over in a complete system of transcendental philosophy, but which in a merely critical essay I can be content with merely mentioning.[63]

In the same sense in his *Advances in Metaphysics,* he says:

> The predicables also belong to the categories as fundamental concepts of the intellect, since they come from the combination of these concepts and are therefore derived from them, either a priori concepts of intellect or sensibly conditioned concepts . . . which could be completely tallied and presented systematically in a table.[64]

We now want to look at just one part of the system of pure reason, namely, that of pure concepts, which is divided into systems of basic concepts and of derived concepts. Kant's usage in describing the division of the system is not entirely fixed so I will keep to the one predominantly used, which is explicitly defined in the *Advances in Metaphysics.*

> Ontology is the part of knowledge (as part of metaphysics) which constitutes a system of all the concepts of the intellect and principles, but only insofar as they apply to objects given to the senses and therefore can be verified by experience. It does not deal with the supersensible, which is really the ultimate goal of metaphysics. It thus is part of metaphysics only as a propaedeutic, as the lobby or outer courtyard of real metaphysics, and is called transcendental philosophy because it contains the conditions and primary elements of all our a priori knowledge.[65]

To bring out even more clearly the aspect which interests us here, I will speak of a *systematic ontology.* Our question can then be put: Is a systematic ontology possible on Kantian principles? Since we want to let the system of basic concepts, the table of categories, stand as valid without dispute, we can also make the question concrete: Do the combined concepts make a complete, definitive system which can be presented exhaustively? Thus we consider systematic ontology in the sense of the definition already quoted from the *Advances in Metaphysics*. It is also identified in other places by Kant as the metaphysics of nature.[66] On the other hand, the meaning of *physica rationalis* vacillates in Kant.

This systematic ontology has to provide the foundations for natural science as well as for mathematics. The transcendental aesthetic and the transcendental analytic belong to it, according to the *Critique of Pure Reason* as well as the *Metaphysical Foundations of Natural Science* and the *Opus Postumum*. We want to ask, therefore, whether this systematic ontology is at all possible as systematic knowledge, and will use a remark of Paul Natorp's as a hint:

> After all this, in the system of the Kantian transcendental philosophy, putting time and space prior to the laws of thinking of objects is a serious mistake which is, at best, understandable and excusable only as anticipation. In a stricter construction of system, time and space would really have to be put under Modality in the category of Actuality; but also in Possibility and Necessity.[67]

On the one hand, we cannot doubt that singling out space and time as pure perceptions presents a basic principle of the Kantian system, but on the other, it is impossible to dispute the work that the Marburg school has done on Kant emphasizing the systematic part. To be sure, it seems quite remarkable that time and space are found to be a peculiar fault of the system. And what if it really were so?—if space and time really disturbed the structure of Kant's system, perhaps even destroyed it? It really could be that Kant's goal, the definitive system of pure philosophy, has been blown to smithereens by his own great discovery, namely, that of the perceptual nature of space and time.

To get closer to this problem we will try to apply Kant's own distinctions to his system, taking it as restricted to systematic ontology. Are judgments of systematic ontology analytic or synthetic? We have to clear up two terminological difficulties. First, the opposition of analytic-synthetic is used in two senses. When we go back to Lambert, the structure is clear enough. He analyzes the concepts given him into their basic concepts—analysis—and then constructs complex concepts from these basic concepts—synthesis. This was the meaning originally intended by Kant and steadfastly maintained. In the 1764 *Enquiry into the Clarity of the Principles of Natural Theology and Morals,* he says:

> It is not yet the time to proceed synthetically in metaphysics; only when analysis has helped us to obtain clearly and fully understood concepts will synthesis then be able to subordinate the simplest knowledge to the complex, as in mathematics.[68]

In this sense, that part of the *Critique of Pure Reason* which presents the system of basic concepts and principles by analyzing the intellect is called *transcendental analytic,* and the part—still missing from the *Critique*—which in order to complete the system constructs the derived concepts from the basic ones and the theorems from the basic principles, must be called the *synthetic.* However, the new distinction, which Kant himself had made, goes in an entirely different direction. There *analysis* is called knowledge which is based solely on the Law of Contradiction and *synthesis* is simply knowledge which goes beyond this.[69] A certain vacillation is also found in the use of the word *metaphysics.* Sometimes it means the whole system of philosophy, so that it includes systematic ontology, while at other times the full meaning of metaphysics is retained as knowledge which goes beyond the bounds of experience, whether to be valued positively or negatively.

Undoubtedly systematic ontology is synthetic knowledge in the special Kantian sense of the term, pure natural science being part of this systematic ontology. The synthetic character of pure natural science is explicitly stated in the *Critique* (B17), as well as in the *Prolegomena* (secs. 14–38). What actual knowledge does pure natural science, and with it systematic ontology, include? The *Critique* (A82 = B108) identifies the following complex concepts which would belong here: Force, Action, Passion, Presence, Resistance, Coming to Be, Passing Away, Change. In the *Metaphysical Foundations of Natural Science,* for example, the parallelogram of forces and Newton's law that the force of gravity is inversely proportional to the square of the distance are deduced a priori. In the *Opus Postumum* Kant wanted to derive even more specific natural phenomena a priori.

Now philosophy, and in particular systematic ontology, is expected to be a science constructed from pure concepts. But how is this supposed to be possible? Can the parallelogram of forces or the law of gravity be deduced from pure concepts? Do not perceptions, in fact, time and space, enter into such laws? And does not the same also hold true for force, resistance, change? If we really wanted to be serious about a science constructed from pure concepts and wanted to get rid of all concepts and principles from systematic ontology into which factors of perception enter, what would be left over then? If we consider the significance of time and space for pure natural science, there can be no doubt that Kant was right in defining it as pure synthetic knowledge. Contrariwise, however, as a part of philosophy it must be a science constructed from pure concepts. But then its propositions would have to be analytic, and indeed they are defined that way by Kant. Kant says explicitly that the *Critique* has already given the synthetic foundations and that the completion of the system is merely analytic labor:

The reason this critique is not itself called transcendental philosophy is simply that to be a complete system, it would also have to contain an exhaustive analysis [*Analysis*] of the whole of human knowledge a priori. Our critique must, to be sure, display a complete enumeration of all the root concepts that make up such pure knowledge. But it need not analyze these concepts exhaustively nor do a complete review of those derived from them, partly because this analysis would serve no purpose, since it is not beset with doubts like those encountered in synthesis [*Synthesis*], for whose sake the whole critique really exists; and partly because it would contradict the unity of the plan to take responsibility for such an analysis and derivation when, in view of the aim of the critique, we can be absolved of it. It is easy to complete the analysis and the derivation of the concepts to be furnished later a priori, if they are present from the very first as exhaustive principles of synthesis, and if they lack nothing necessary for this fundamental aim.[70]

We may use *synthesis* in this section to refer to whatever we want, but it is still evident that the synthetic part of the task has basically been done. What is missing is called *analysis* and *derivation* by Kant.

By using these two terms very carefully, Kant avoids the difficulty we pose: a difficulty which really lies in the fact that systematic ontology is supposed to be knowledge from pure concepts so that its propositions should be analytic, even though it is only too easy to see that the pure perceptions—of time and space—are indispensable to it, and that *this* shows its propositions are synthetic.

How would such a system of derived concepts be possible at all? We go back once more to Lambert, who in this case only continued Leibniz's train of thought. Lambert assumes that the system of basic concepts has already been discovered, or at least that those known to us are part of this system, so that we can work out all the combinations. If a contradiction results, then the complex concept is rejected as impossible, but if it is free from contradiction, then it is possible and the concept so constructed is real. In this sense, therefore, knowledge of the derived concepts would be a pure combinatoric based solely on the Law of Contradiction. Similar considerations may also have led Kant to identify such knowledge as analytic. On the other hand, as he indicates, the concepts free of contradiction must be divided into those which are empty and those having content. So another law of gravity, say according to the cubic power, would be conceivable entirely free of contradiction. Now since Kant derives Newton's law a priori, a foundation with empirical factors is excluded and the special feature of the second power can only be based on the pure perceptions. If this holds true for the whole of systematic ontology, then it is, indeed, related to perception and is thus synthetic.

What would the method of a systematic ontology look like? We have seen that

Kant requires an *ars characteristica universalis* to create the system of derived concepts, that he even brought a transcendental analysis and a transcendental algorithm into his speculations.[71] But purely mathematical methods would be introduced into philosophy thereby and ontology would thus require a symbolic construction. In fact, Kant even says the following in the *Metaphysical Foundations of Natural Science:*

> I have nevertheless, followed the mathematical method in this treatise, although I have not adhered to it with complete strictness (more time would have been required for it than I had), not in order to give it a better introduction by a flashy display of profundity, but because I believe that such a system is really possible and that this perfection can well be attained in time by a more skilled hand.[72]

Kant says clearly at this point that the metaphysical foundations of natural science should be dealt with by purely mathematical methods. But now he expressly states that this work is supposed to be a part of the promised system of pure reason. Therefore the system of pure reason really has to be dealt with, at least in ontology, by purely mathematical methods; but then what remains of the difference between philosophy and mathematics which Kant argued for so vigorously?

At least in arithmetic, the capacity for expansion to infinity arises as the third characteristic of synthetic branches of knowledge. From this standpoint as well, however, difficulties arise for systematic ontology. The term *expanding* [*erweiternd*] could have a threefold meaning for Kant: (1) A new discovery expands knowledge when it does not consist of tautological propositions but leads to a real increase of knowledge. (2) Such an increase could, however, be unlimited; then this science would be called self-expanding in the strict sense of the word as, for example, arithmetic. (3) Science also expands when it extends beyond the boundaries of possible experience.

We have restricted ourselves to ontology and thus do not have to take the third meaning into account. The question is only whether ontology leads to a limited amount of knowledge or whether it is a science which increases without limit. The first follows from all that Kant says in the *Critique of Pure Reason* about the possibility of a complete system, and specific declarations are also found elsewhere: In addition to the *Advances in Metaphysics*, for example, the *Metaphysical Foundations of Natural Science* says "that the absolute completeness of knowledge can be hoped for in whatever is called metaphysics."[73] So here also the completeness of the metaphysics of corporeal nature can be expected with confidence.

On the other hand, Kant defines systematic ontology as a real expansion of

our knowledge, about which there can really be no doubt anyway. In the
Advances in Metaphysics, he says:

> The expansion of knowledge a priori, by pure concepts, also outside of
> mathematics, and that it [this expansion] contains truth, is shown by the
> agreement of such judgments and basic principles with experience.[74]
>
> For what reason says about the expansion a priori of the knowledge of
> objects of possible experience in mathematics as well as ontology are real
> steps forward by which it is sure to gain some ground.[75]

If we admit that ontology is a field of expanding knowledge, and if, contrariwise, we really suppose it possible to confidently expect completeness from it,
then there must be a limit to which ontology as a priori knowledge is possible. By
reaching it, ontology comes at the same time to a definitive conclusion. Nowhere in
Kant, however, is there even a hint of such a limit nor even a hint of what such a limit
would look like. Force, Resistance, Coming to Be, Passing Away, Change are
supposed to be concepts belonging to a priori ontology. But then why not also
Melting and Freezing? Why not Crystallizing and Dissolving? Where is the limit to
the expansion supposed to be? It is no different with the laws. The parallelogram of
forces and Newton's law of gravity are supposed to be deducible a priori as part of
systematic ontology. Then why should not a law relating pressure to temperature be
a priori? Where is the limit supposed to be?

We cannot overlook the fact that the unlimited apriorizing of German idealism has its origin here. True, on the one hand, Kant did not really take this step to
the theory that all knowledge is a priori without limit. Still, the German idealists
did clear the way to it only after abandoning pure perceptions as significant.
Thus we begin to conclude that any special significance given to pure perception is incompatible with the idea of a completed system.

The difficulties stem chiefly from the distinction between philosophy and
mathematics which Kant always maintained. See, for example, the *Critique of
Pure Reason:* "Philosophical knowledge is rational knowledge from concepts,
mathematical from construction of concepts."[76] He repeats this in the *Metaphysical Foundations of Natural Science:*

> Pure rational knowledge from mere concepts is called pure philosophy or
> metaphysics; as opposed to this, knowledge which is based only on the
> construction of concepts by means of the representation of objects in
> perception a priori is called mathematics.[77]

According to this distinction, there is no way to classify systematic ontology
as the metaphysics of nature. The distinction could, indeed, be applied since the
metaphysics of nature also deals with pure perceptions, insofar as—merely

according to pure concepts—they are meaningful for this discipline. But Kant has cut himself off from such a distinction by his explicit account that the metaphysical foundations of natural science must be handled by purely mathematical methods. But the metaphysics of nature is not the only discipline whose position in the system makes difficulties. For example, the question of where to put the analytic principles exhaustively is not at all easy to answer. The greatest difficulty, however, is presented by time itself. Is there any a priori knowledge of time at all? With space, the problem appears to be easy; geometry is the a priori knowledge of space. Yet even here, the complication also arises that geometry (the theory of a particular kind of quantum) is already regarded by Kant as a special case of general mathematics (as the theory of quantity [*Quantitas*]).[78] Thus, an entirely different kind of knowledge lies at the base of geometry, namely, pure mathematics. For time, there are three possibilities:

1) There is a special a priori science of time.

2) General mathematics [*Mathesis*], and thus arithmetic, algebra, and analysis with a purely arithmetic foundation, is the theory of time.

3) The pure theory of motion is the theory of time.

There is no support in Kant for an independent science of time. We have just seen that in the transcendental aesthetic, arithmetic is not referred to as the theory of time. On the contrary, a remark is inserted into the second edition which carefully establishes a connection between time and the general theory of motion: "Thus our concept of time explains the possibility of as many synthetic sciences as the general theory of motion, no less fruitful, presents."[79] A similar observation is made about geometry: "Our explanation is thus the only one which makes intelligible the possibility of geometry as synthetic knowledge a priori."[80]

But the general theory of motion cannot possibly be the pure science of time since spatial determination always enters into this science. This problem can also be seen in Kant's pupils. In a remark about geometry, Schultz rejects any special science of time; for him, this science is identical with the geometry of the straight line and because of its limited range does not deserve to be treated as a special science.[81] Kiesewetter, on the contrary, remains true to the Kantian approach and requires a science of time peculiar to itself, without being able to find it.

All these obstacles which the pure perceptions put in the way of completing a system can be summed up when we look at the advice Kant himself gives for such completion:

The categories, when combined with the modes of pure sensibility or with one another, give a great number of derivative concepts a priori. To note and, where possible, to list them completely would be a useful and not unpleasant task but one that can be dispensed with here.[82]

The question whether, according to this, a complete table of the a priori concepts is to be set up has to be answered with a flat *no*. Anyway, it is still noteworthy that in this passage Kant clearly expresses doubts about such a result by inserting "where possible," while in other places he assumes there is no doubt that completeness can be achieved. To be sure, it would be easy to begin by combining the pure concepts and so the categories according to these guidelines; it is true that completeness in the derived concepts could only be expected here when a limit is set to the number of combinations.

We would have to stipulate that a complex concept could contain no more than twelve parts. If, on the contrary, there were no such restriction, then the combining, and thus the number of derived concepts as well, would go on ad infinitum. The other part of the task, that of arriving at a complete table of a priori concepts by combining the categories with the modes of pure sensibility is, however, entirely insoluble. We can see at once that for completeness, not only the categories but also the modes of pure sensibility must be finite and determinate in number. For Kant this holds for the categories: There are exactly twelve. According to Kant himself, however, such a systematic enumeration is wholly unthinkable for the modes of pure sensibility; rather, the Kantian approach holds that pure sensibility has to simply be accepted by us human beings. We do not even know why we have only two pure perceptions and why we have these particular ones. Thus, not only did Kant himself give no exact enumeration of the modes of pure sensibility, but it would also be intrinsically impossible to do so. For even if particular modes of pure sensibility are counted, it could never be proved that the enumeration was complete since, according to Kant, one cannot deduce pure sensibility.

We must therefore conclude that the part which the pure perceptions contribute to the system of pure reason cannot be definitively determined. But then the idea of such a system as a complete structure to be finally executed is exploded. And so not only is Paul Natorp's observation shown to be right: but also the special role of the pure perceptions is, in fact, incompatible with the idea of a finished system; the difficulties dealt with just now all come from the same source. The pure perceptions, particularly that of time, manifest themselves as permanent obstructions in the system. And perhaps Kant's greatness lies in the fact that he did not force a solution of these problems and erect a completed system, but that instead he pursued these questions as far as possible and then left them there with all the question marks strewn around them.

CHAPTER SIX

Synthetic Judgment in Arithmetic

ALTHOUGH THIS CHAPTER BEGINS with some interpretation, we must admit that really penetrating interpretations will be possible only when we can have arithmetic at our disposal. In these questions, too, then, the present work will confine itself to presenting materials.

A. *The Natural Numbers*

I. REFERENCES

Kant deals with number in different places in the *Critique of Pure Reason*. This has given rise to difficulties and mistakes. In his Dissertation of 1770, number had already been referred to in the chapter corresponding to the transcendental aesthetic:

> In addition to these concepts there is another one which in itself is indeed intellectual but whose realization in the concrete requires the aid of the notions of time and space (by successively adding a number of things and by simultaneously putting them next to each other). This is the concept of *number*, which is dealt with in Arithmetic.[1]

The transcendental aesthetic does not deal with number, although this is continually maintained.[2] The relevant explication in Kant is to be found in three places in the *Critique*: in section 10 of the exposition of the table of categories in the context of the theory of pure rational concepts, in the deduction of pure rational concepts, and in the analytic of principles in the chapter about schematism.

Kant speaks about number twice in the *Critique* in the context of the theory of pure rational concepts:

> Pure *synthesis, thought of generally*, gives us the pure concept of the intellect. By this synthesis, however, I mean that which depends on a base of synthetic unity a priori. Thus our counting (above all seen in the larger

numbers) is a *synthesis according to concepts*, because it is done according to a common base of unity (e.g., the decimal).[3]

Paragraphs 11 and 12 were added to this section 10 in the second edition for further clarification. Kant says there:

> So *Allness* (Totality) is just plurality considered as unity. . . . So the concept of a number (which belongs to the category of Allness) is not always possible where the concepts of plurality and unity are present (e.g., in the idea of the infinite).[4]

In the transcendental deduction Kant illustrates the levels of synthesis by reference to number. In the following, we will consider these passages more closely. They are found only in the first edition.

The chapter on schematism explicitly states what numbers are:

> The pure image of all quantities (*quantorum*) for outer sense is space; that of all objects of the senses in general is time. But the pure *schema* of *quantity* (*quantitatis*), as an intellectual concept, is *number*, an idea which comprehends the successive addition of one to one (homogeneous). Therefore, number is simply the unity of the synthesis of the manyfold[5] of a homogeneous perception in general, due to my generating time itself in apprehending the perception.[6]

And in the second passage:

> But it is indeed also striking that although the schemata of sensibility first and foremost serve to realize the categories, at the same time they restrict them, that is, limit them to conditions outside the intellect (namely, in sensibility). So the schema is really only the phenomenon, or sensuous concept of an object, which agrees with the categories. (*Numerus est quantitas phaenomenon, sensatio realitas phaenomenon, constans et perdurabile rerum substantia phaenomenon—aeternitas necessitas phaenomenon*, etc.).[7]

If we want a concise definition at all for such a many-faceted phenomenon as number, the easiest to take as the decisive Kantian definition is this one: Number is phenomenal quantity [*Numerus est quantitas phaenomenon*]. The *Crtique of Judgment* returns to number without saying anything essentially new.[8]

2. THE PROBLEMS

Kant has been continually criticized for obscurity and vagueness in the discussion of these questions. The first dissertation about Kant's theory of analytic and synthetic judgment, Carl Schulze's *Dissertatio Inauguralis exhibens nonulla ad doctrinam judiciis analyticis atque syntheticis spetantia*, which appeared in 1793, complained: "What kind of thing is our idea of number? Is it a perception? or a concept?" [*Qualis est nostra number repraesentatio? Estne intuitus? an conceptus?*][9]

Michaelis, in his 1884 article about Kant's concept of number, also complains about vacillation, saying that Kant expelled "the concept of number totally from the sphere of sensibility into the realm of intellect" in the passages following the table of categories.[10] In contrast, the discussion in the chapter on schematism refers to the concept of number as inseparable from the perception of time. Michaelis sees an unresolvable contradiction between these two positions.[11]

The contradictions are, in fact, not as great as Michaelis finds them. Moreover, *number* is such a difficult concept that any facile solution must, of course, arouse suspicion from the first. What, then, are the basic questions an investigation of number has to answer?

First of all, it is maintained that on the one hand the concepts of mathematics, particularly those of number, exist in themselves and that mathematical propositions are eternal truths also existing in themselves. The other side argues against this, denying that there is mathematics independent of man. A further question makes sense only in the context of the Kantian assumptions: Is number a pure concept or a perception? Two further questions, concerning the arithmetization and logicizing of number, were more clearly developed in the nineteenth century.[12]

Let us elaborate on the first question; two possibilities are often posed in asking: Are new mathematical propositions discovered or invented? I will briefly refer to Bolzano. He considers ideas-in-themselves which exist independently of whether anyone thinks of them.

> What does the author mean by a proposition? . . . by a proposition-in-itself I mean only any assertion that something is or is not; no matter whether this assertion is true or false; whether anyone expresses it in words or not, or even whether any minds think of it or not.[13]

Furthermore, according to Bolzano, although this does not determine what a truth-in-itself is, all ideas and truths-in-themselves and so, in particular, all mathematical concepts and judgments, are eternally imagined and thought of by God who is omniscient. Beginning in this way, Bolzano has undoubtedly expressed the real kernel of the Leibnizian theory. For Leibniz as well, mathematical judgments are eternal truths [*veritates aeternae*] which, in particular, are eternally known by God. This position is so self-evident for Leibniz that he in fact speaks of a divine mathematics [*mathesis divina*]: "From this it is amazing to realize that at the very beginning of things some divine mathematics or metaphysical mechanism is used and a maximum limit is possible."[14]

We see here that Kant hardly follows Leibniz. For Kant, indeed, the difference between an infinite mind and the finite mind of men is really so immeasurably great that it is hardly to be expected that they could both have the same mathematics.

3. KANT'S CONSTRUCTION OF THE NUMBERS

We may begin with the transcendental deduction of the first edition of the *Critique of Pure Reason*, where Kant uses the concepts of time, straight line, and number as examples of the three subjective sources of knowledge.

Kant distinguishes the synthesis of apprehension [*Apprehension*] in perception, the synthesis of reproduction [*Reproduktion*] in imagination, and the synthesis of recognition [*Rekognition*] in concepts:

> Now it is obvious that when I mentally draw a line, or want to think of the time from one noon to another, or of a particular number, I must first grasp [*fassen*] these various ideas one after the other. If I were, however, always losing the preceding ones (the first part of the line, the preceding parts of time or the preceding units which come in succession) and not reproducing them while I proceeded to those following, no complete idea . . . , not even the purest and most elementary ideas of space and time could ever arise.[15]

This train of thought is continued in the "Synthesis of Recognition in Concepts" as follows:

> If in counting I forgot that I had added the units I was thinking of one after the other in succession, I would never know that a whole is being produced by this successive addition of one to one, and so would not recognize the number. . . . For it is one consciousness which unites the manyfold, which has been perceived in succession and then reproduced, into one idea.[16]

I will give quite a simple instance of constructing a number. When truck drivers want to unload a truck full of sacks of potatoes, they usually mark off each sack brought in on a stone at the entrance to the building with a piece of chalk. Every fifth sack, however, is marked off not with a straight vertical line but with a diagonal, so that it looks like this:

Twenty-three sacks will then look like this:

In this simple procedure of counting, the three stages indicated by Kant can be distinguished very clearly. First, the plurality of the sacks is seen by noting them—piece by piece as each is carried in. Since the man does not want to count in his head and cannot do it while he is working, he notes each sack with a line. Further counting can be done using the lines. The second requirement for counting is that the chalkmarks which are now being counted instead of the sacks should be put in a row.

If the truck driver were to put the mark here at one time and there at another, he would know no more afterward than before. In a certain sense, a continuing row of chalk marks could already represent a number.[17] Leopold Kronecker (1823–1891) said as much in his 1887 article on the concept of number: "But if repetition is permitted, then a single symbol is enough to express any number because the symbol is repeated as often as indicated by the number."[18]

But this kind of counting is impractical even for the truck driver; rather, he introduces the true conceptualization of number when he creates a pentadic system by means of the fifth line drawn as a diagonal. He makes do with it for numbers necessary for truck drivers since he can at a single glance grasp pretty well all the numbers up to fifty. If the truck drivers had needed larger numbers, they would probably have invented a new unit containing a new combination of, say, five groups of five. Thus the Kantian steps are made clear. To count means to go through different things, item by item, make them reproducible—in counting, by the arrangement into a series—and putting what is reproduced together by means of a concept so it can be represented as an integrated idea.

Kant has established a variety of points here. In counting, a group of different things is always counted as a series. In the beginning, it will always consist of things which are somehow needed. Later, the counting can be freed from these, and then the symbols and even the concepts themselves can also be counted, and finally—to apply it to itself—symbols of the numbers and symbols of the concepts can be counted as well.

For Kant the division of the cognitive faculties into sense, imagination, and apperception is intended as an analysis of the process of counting as consisting of *going through* [*durchgehen*] (the different things perceived), *taking* [*them*] *together* [*Zusammennehmen*] to form a series which can be reproduced, and *combining* [*Zusammenfassen*] them into the unity of the concept. This division is to be found not only in the transcendental deduction of the first edition of the

Critique but is consistently made by Kant, as, for example, in the following passage from the section about the pure concepts of intellect, or the categories, in both editions: "But if we are to recognize this manyfold as such, the spontaneity of our thinking requires that it first be gone through in a particular way, then taken up and united."[19] The chapter on the schematism is in accord with this: "Therefore, number is simply the *unity* of the *synthesis* of the *manyfold of a homogeneous perception in general.*"[20] The *Critique of Judgment* also distinguishes the assembling by the imagination from the summing up in the concept, whether decimal or tetradic.[21]

In line with this, the following subsidiary problems arise. Schultz emphasizes in his *Anfansgründe* that according to Kant, number always presupposes something countable:

> It is clear that the concept of a whole number can be arrived at only . . . through a comparison of a quantity with another homogeneous with it, taken one or more times. For counting without things to think about which are counted is absurd. So we only arrive at the concept of all the rest of the numbers as well as of the number 1 in the following way.[22]

The construction of a reproducible series is, furthermore, contained in counting. It seems as if the approach to set theory lies here and as if set theory comprehends the numbers as reproducible (but comparable) series, in order to obtain the basic concept of equivalence.

For Kant it is apperception which really gives number the property of being a concept and this property manifests itself in the decimal system, though any number can serve as the base of the system. The problems of the decimal system were discussed in various ways in the seventeenth and eighteenth centuries. It had already been seen early on that the base of the number system is really arbitrary. This fact takes on special significance since the multiplication table depends on it. The basic operations to be performed by the basic numbers of the system—in the decimal system the numerals 1 through 9—cannot be calculated but must be gotten through perception and must be learned by heart, like the so-called multiplication table. Newton stated this explicitly at the beginning of his *Arithmetica universalis*:

> On addition: The addition of numbers which are not completely composite is in itself obvious. So 7 plus 9 or 7 + 9 obviously makes 16, and 11 + 15 obviously makes 26. . . .
>
> On multiplication: Numbers which come from multiplying any two numbers not greater than 9 ought to be learned by heart.[23]

Leibniz said that a difficult problem was hidden here in this "obviously." He was also the first to see that there is no multiplication table in a dyadic system, and he therefore believed that he could overcome all the deeply rooted difficulties here by developing the dyadic system: "[T]he binary number system is intrinsically superior to the decimal or any other, since in the binary system everything which is asserted about numbers can be shown from the symbols. This is not true of the decimal system."[24] Wolff praised him for this:

> [T]he admirable Leibniz also discovered a binary or dyadic arithmetic which does not go beyond two numbers and can be used by the learned to investigate the hidden properties of the numbers, since it resolves them into the primary elements, o and 1.[25]

In the 1771 *Architektonik* Lambert goes into the questions of constructing numbers in considerable detail, also referring to the special significance of the dyadic system for Leibniz. Lambert had already indicated that there are properties of numbers independent of the number system, for example, the property of a number's being prime. Although Kant in dealing with numbers uses very careful and comprehensive formulations of the unity of apperception in a concept, here we can really see his effort to include other concepts, for example, prime number, in the class of such arithmetical concepts as that of the base of the decimal system.

Furthermore, we see that the decimal numbers are in no way homogeneous; they fall into three groups, the first being the number One which has a special position since the threefold process of counting sketched out by Kant is inapplicable here. For the number One there is no perception of different things, *taking* [*them*] *together* by the imagination nor *putting* [*them*] *together* by apperception. The second group consists of the small numbers, those going up to 5 or 10 or 15, that is, numbers which can be illustrated on the fingers simply by looking. The third group consists of the large numbers which can only be constructed symbolically in a system; for example, 2,351 can only be given through symbols. It is here that the problems of symbolic construction enter and these are, according to Kant, the peculiar mark of arithmetic.[26]

It will suffice to say that according to the chapter on schematism, Kant stands in firm opposition to Leibniz. Kant's attempt to find the function and significance of a schema for the overall direction of thought may be obscure or clear; in any case, it is clear that number is bound up with time and thus with a form of the finite mind of man, and it is clear that number is an appearance. Accordingly, mathematics, and particularly number in its totality, belongs to the finite mind and thus fits the description in the *Critique*:

For if I wanted to think of an intellect which itself perceived [*der selbst anschaute*] (like perhaps a divine one [which does not have ideas of concrete objects but whose very having of the ideas presents the objects themselves, or produces them]), the categories would have no meaning whatsoever with respect to such knowledge. They are only rules for an intellect whose entire capacity consists in thinking, [i.e., in the activity of bringing the synthesis of the manyfold . . . to the unity of apperception].[27]

This account is again given expressly for mathematics shortly afterward: "It follows that no mathematical concept is in itself knowledge, except on the assumption that there are things which can only be presented to us in the form of that pure perception of sense."[28]

4. THE ARITHMETIZATION OF NUMBERS

In his essay on the concept of number Leopold Kronecker interprets the arithmetization of mathematics to be the founding of the whole of mathematics on the concept of number.[29] Such an arithmetization of mathematics presupposes, of course, that the pure concept of number is grasped independent of all applications—in particular of the geometrical meaning with which particular classes of numbers may have first appeared. It is well-known that the Greeks saw the irrational numbers primarily as a geometrical problem. These numbers first appeared in the diagonal of a square and offered much resistance to an arithmetical interpretation. We have seen above (p. 58) that even Wolff describes the irrational numbers purely geometrically, namely, as the relation between one straight line and another. These problems were divided into two different tasks in the seventeenth and eighteenth centuries. On the one hand, the irrational numbers had to be separated from their geometrical applications; on the other, from differential and integral calculus, originally discovered purely geometrically just like the irrational numbers. It was only the solution in terms of the concept of limit that made the interrelation clear. Euler was probably the first to consider a purely arithmetical foundation for differential and integral calculus, yet the textbooks of the time show how little the mathematicians had become generally conscious of the necessity for this. I cite, by way of example, Abraham Kästner's *Anfangsgründe der Arithmetik, Geometrie, ebenen und sphärischen Trigonometrie und Perspective* of 1792: "All the sciences which belong to analysis (algebra, differential and integral calculus) are really large chapters of arithmetic or geometry or of a science composed of both of them."[30]

Schultz, on the other hand, is entirely clear about the fact that arithmetic, and in particular the irrational numbers and the infinitesimal calculus, must be isolated from all geometrical considerations. Accordingly, in contrast to Wolff, in the *Anfangsgründe* he defines irrational numbers without any reference to geometry, making a similar claim for the infinitesimal calculus:

When two homogeneous quantities *a* and *b* can be produced by some common measure, they are called commensurable or rational quantities. If, on the contrary, they cannot be produced by any common measure that may be assumed, they are called incommensurable or irrational quantities.[31]

What is called general mathematics (*mathesis universalis*) is that part of pure mathematics which investigates in general the possible combinations of what is homogeneous by which a quantity can be produced, and can thus be entirely abstracted from the qualities of the quantities. On the other hand, that which has as its object a quantity of a particular quality is called special mathematics (*mathesis specialis*). The former is thus the basis of this as well as of the whole of applied mathematics. . . .

Since algebra together with differential and integral calculus is simply higher arithmetic and thus belongs solely to general mathematics, these three fields, as well as arithmetic, must be handled entirely divorced from special mathematics. It is, therefore, a change to another genus [*metabasis eis allo genos*] when the proofs of their theorems are taken from geometry or trigonometry. On the whole, the entire system of general mathematics, since it is the basis of all special and applied mathematics, must be reasonably established so that even someone who does not know the least thing about geometry is in a position to understand it fully and thoroughly.[32]

This distinction between a fundamental science of mathematics which deals with quantity as such and geometry which, for its part, deals with a particular kind of quantity, namely, that of space, goes back to Kant:

But mathematics does not only construct quantities (*quanta*) as in geometry; it also constructs quantity as such (*quantitatem*), as in algebra where it completely abstracts from the quality of the object according to which such a concept of quantity is to be thought.[33]

The pure image of all quantities (*quantorum*) for outer sense is space; that of all objects of the senses in general is time. But the pure *schema* of *quantity* (*quantitatis*) as a concept of the intellect is number.[34]

5. THE LOGICIZING OF NUMBER

When we demand a purely arithmetic representation and foundation of numbers, we must obviously ask whether it is not possible to give a purely logical representation of numbers. Here, of course, we ought not to get stuck in terminological issues but must deal with the substantive matters. If we mean by *logic* something entirely different from what Kant meant, then it is, of course, easy to oppose Kant. The original author of the idea of a purely logical foundation of mathematics is indisputably Leibniz, whose views I have already to some extent compared to Kant's. I will go into them some more later in examining concrete, individual questions, since, to my mind, that is the only fruitful approach.

This approach to a purely logical foundation of mathematics is to be found in three philosophical schools, those of Marburg, of phenomenology and of the Vienna Circle (Rudolph Carnap, Hans Reichenbach). For example, I will cite Paul Natorp here:

> The distinguishing characteristic lies in this: whether the basic concepts of a science come from logic, whether they are likewise concepts of logic, or whether its primary axioms are contained in the primary laws of logic or can be derived from them.[35]

(As we have seen, Leibniz had already demanded the proof of all the axioms.)

B. Arithmetical Judgment
I. REFERENCES

We turn now to a careful investigation of Kant's results about arithmetic judgment. The first edition of the *Critique* had already contained the statement "$7 + 5 = 12$ is not an analytic proposition":

> $7 + 5 = 12$ is not an analytic proposition. For I do not think of the number 12 in either the idea of 7 or in that of 5, or in the combination of both. (The point here is not that I have to do this in the *addition of the two*; for with the analytic proposition the question is only whether I actually think of the predicate in the idea of the subject.)[36]

The main passage is in the *Prolegomena*, and is taken over into the introduction of the second edition of the *Critique* with an important addendum (*Prolegomena*, Ak. IV, 268ff. and the *Critique*, B15ff.):

> We might, indeed, at first think that the proposition $7 + 5 = 12$ is a merely analytic proposition and follows by the law of contradiction from the concept of a sum of seven and five. But if we look more closely we find that the concept of the sum of 7 and 5 contains nothing but the union of the two numbers into a single one, by which nothing at all is thought about what this single number is which combines the two. The concept of 12 is in no way already thought of when I merely think of this union of seven and five; and I may analyze my concept of such a possible sum as long as I please, but I will never find the twelve in it. We have to go outside these concepts and use as help the perception which corresponds to one of them, our five fingers, for instance, or (as Segner does in his *Arithmetik*) five points, and so add one by one the units of the five given in perception to the concept of seven.

Prolegomena, 269	*Critique*, B15ff.
Thus we really enlarge our concept in the proposition $7 + 5 = 12$ and add a new concept to the first which was not thought of at all in it,	For I first take the number 7 and, using the perception of the fingers of my hand for help for the concept of 5, I now put the units which I had taken to make that image of mine up to 7, and so I see the number 12 arise. I have, indeed, thought that 7 and 5 are to be added when I have the concept of a sum $= 7 + 5$, but not that this sum is the same as the number 12.
i.e., the arithmetical proposition is, always synthetic,	The arithmetical proposition is, therefore, always synthetic;

[Both texts]
which is still more evident if we take larger numbers. For then it is obvious that by turning and twisting our concepts, as much as we please, we could never find the sum by merely analyzing our concepts without the aid of perception.

(The second and most important reference, which is like this basic source, is the long letter of Kant to Schultz in 1778 that I have already described in Chapter 3. I will use it later for further interpretation.)[37]

The philosophical discussion was begun by Locke, whose *Essay Concerning Human Understanding* deals with the axioms in Book IV, chapter 7, section 10:

And indeed, I think, I may ask these men, who will needs have all knowledge, besides those general principles themselves, to depend on general, innate, and self-evident principles. What principle is requisite to prove that one and one are two, that two and two are four, that three times two are six? Which being known without any proof, do evince, That either all knowledge does not depend on certain *praecognita* or general maxims, called principles; or else that these are principles: and if these are to be counted principles, a great part of numeration will be so.[38]

Then Leibniz took up the question in his *New Essays*, likewise Book IV, chapter 7, section 10, and gave the following proof:

[Theophilus] That two and two is four is not a completely immediate truth, namely, when four means three and one. We can demonstrate this, and here is how:

Definitions

1. Two is one and one,
2. Three is two and one,
3. Four is three and one.

Axiom

If an equal be substituted for an equal, then the equality remains.

Proof

2 and 2 is 2 and 1 and 1 (according to Def. 1),	$2 + 2$
2 and 1 and 1 is 3 and 1 (according to Def. 2),	$2 + 1 + 1$
3 and 1 is 4 (according to Def. 3).	$3 + 1$
	4

Therefore (according to the axiom) 2 and 2 = 4. Q. E. D.

I could also have said that 2 and 2 is equal to 2 and 1 and 1 instead of saying that 2 and 2 is 2 and 1 and 1, and similarly with the rest. But we can assume that this has already been done throughout on the strength of another axiom that anything is equal to itself, that whatever is the same is equal.[39]

Today it is generally conceded that this proof of Leibniz's is inadequate because it uses the law of association.[40] But I want to get at Leibniz's basic position in the strict sense. If we go back to the Leibnizian distinction between complex and simple concepts, then the concept of four is unquestionably complex, made up of components. The concept *homo* is the concept *animal rationale*. In this sense, the concept of four is made up of the two concepts of three and one. Thus to be precise, we should not write "four equals three + one," which Leibniz also, to be sure, does not write, but instead ":four: is :three, one:".

Since three can also be resolved into components, four analyses of the complex concept of four into component concepts result when their arrangement is not taken into account. Thus we have

> :four: is   :three, one:
>
> :two, two:
>
> :two, one, one:
>
> :one, one, one, one:

Two considerations show that this is really the right interpretation. Leibniz himself says in the passage cited that four is not only *equal* to two and two but also that it is simply two and two. The complex concept of four must, therefore, be just as much the sum of its two component concepts, *two, two,* as the concept of *homo* is the sum of its component concepts, *animal* and *rationale*. Furthermore, the decomposition of a complex concept into its parts presupposes differences of degree in these analyses. That is, there must be an ultimate analysis in which the complex concept is really decomposed into simple parts. Afterwards, fewer primary analyses appear, analyses in which complex concepts are still used. Leibniz did, in fact, draw this conclusion:

> For example, the property of the number Ten of being equal to $6 + 4$ is posterior to that of being equal to $6 + 3 + 1$ since the second expression is closer to the original definition of the number 10, by which it equals $1 + 1 + 1 + 1 + 1 + 1 + 1 + 1 + 1 + 1$. Still, it can be conceived without the second property, and what is more, it can be proved without it.[41]

2. INTERPRETATION OF $7 + 5 = 12$

Interpretations of Kant's views about arithmetical judgment frequently go wrong because they stick much too closely to Kant's nominal definition of synthetic and analytic judgments. Continual attempts then follow to discover contradictions in the concept of analytic judgment. People ask how the concept of the predicate could be contained in that of the subject, how the former could be derived from the latter, and thereby two things are overlooked. First, there are according to Kant no analytic judgments at all in mathematics in the strict sense. If Kant does not state this straightforwardly enough, perhaps the reason is that now and again propositions of the following form are used in mathematics: "The equilateral triangle is a triangle." If, for example, we want to prove a theorem about equilateral triangles, it would be natural, when starting from the proposition "The equilateral triangle is a triangle," to go back to the general proposition, "The sum of the angles of a triangle is two right angles." Although such judgments are not particularly frequent in mathematics, or particularly valuable, there is still no reason for discarding them entirely. Hence, Kant's careful manner of expression can be seen as taking such propositions into account. In addition, we must note that Kant's primary purpose is not to make a logical distinction between two kinds of judgment, the analytic and the synthetic, but rather to characterize the relevant field, in our case that of arithmetic, by synthetic judgment.

The main passage in the *Critique*, B15, is in three sections:

1. The concept of a sum ("But if we look more closely . . . but I will never find the twelve in it)."

2. The addition of small numbers ("We have to go outside these concepts. . . . The arithmetic proposition is therefore always synthetic").

3. The addition of large numbers ("which is still more evident . . . we could never find the sum").

We will begin by establishing what concerns the third section and consider an objection frequently made, as formulated by Hermann Hankel in 1867 in his *Theorie der komplexen Zahlen*:

> the foundation of the number formulas . . . by the five fingers. . . . And if the proposition $2 \times 2 = 4$ can be established in this way, we have to give up the attempt to prove in the same way the proposition that $1000 \times 1000 = 1,000,000$, although Kant recommends the former.[42]

At first glance, Hankel is right. In the second section, Kant talks about the fact that in executing $5 + 7$, the perception of the fingers must be used to help. Kant then says in the third section that the need to use the fingers will be seen much more clearly if somewhat larger numbers are taken. But Kant cannot possibly have meant that, for he certainly knew that people have only ten fingers and therefore cannot do the calculation of $1000 \times 1000 = 1,000,000$ on these ten! But is going from the terms to the sum a purely conceptual conclusion just because there are not enough fingers to do this? Certainly not; rather, with larger numbers the sum is calculated. That means, therefore, that arithmetic has a calculus, or rather the reverse, arithmetic calculation is the model [*Grundtyp*] for any calculus. But any calculation occurs on paper or on a blackboard, i.e., through spatial perception. The rules of arithmetic show that this is not accidental, e.g., those for addition. The numbers to be added must be written one under the other so that the tens stand under the tens, and so on. Without a doubt this way of writing one under the other contains a spatial ordering and thus perception. It can then be seen that, in general, the positional notation of decimal numbers presupposes spatial perception. Mathematical work universally operates with spatial symbols and therefore uses perception. Kant indicated this by ascribing symbolic construction to arithmetic. So far as calculating with large numbers is concerned, there can be no doubt that operating with numerals presupposes spatial perception.

From this we can see what value Kant placed on demonstrating that primitive calculating is done with the help of looking at the fingers. We should not, however, be deceived by this. The core of his thought is that any kind of arithmetic presupposes perceptual aspects in its symbolic construction. Therefore it holds equally well whether primitive calculating uses fingers or more sophisticated calculating works with numbers on paper. Furthermore, in any kind of calculating there are intermediate modes between using fingers and using paper. We maintain that a result of the second and third sections is that arithmetic is bound up with symbols and thus with perception because of its symbolic construction.

In the first section, Kant states that the number twelve is not contained in the concept of seven and five. Many interpretations have been based on this very formulation. If twelve is not contained in the concept of such a sum, we can then turn the question around and ask: what *is* contained in the concept of such a sum? Seven plus five can be formulated in two ways:

$$7 + 5 = 12$$
$$7 + 5 = 5 + 7$$

The first formulation would indicate that the result is not yet in the concept of a sum but must first be calculated in the way indicated. The second formulation would show that the axioms defining the addition of positive integers are not contained in the concept of a sum. Kant wants to say both things. The concept of a sum contains neither the result nor the axiom. The latter claim may appear surprising at first but it makes more sense if one refers to the paragraph immediately following this section, *Prolegomena* 269/*Critique* B16:

> Likewise, no principle of pure geometry is analytic. That the straight line between two points is the shortest is a synthetic proposition. For my concept of *straight* contains nothing about quantity but only about quality. The concept of the shortest, therefore, is completely an addition and cannot be derived from the analysis of the concept of the straight line in any way. Perception must therefore be used here to help; the synthesis is only possible by means of it.[43]

Both these paragraphs, *Prolegomena* 268/*Critique* B15 and *Prolegomena* 269/*Critique* B16, are in accord. Just as in arithmetic the concept of twelve cannot be derived by any analysis of the concept of the sum of seven and five, so in geometry the concept of the shortest cannot be derived by any kind of analysis of the concept of straight. With this position, Kant entered into the eighteenth-century discussion and took the axiomatic approach he had newly developed in considering the question whether there is a concept of *straight line* from which the relevant axiom can be derived. If this interpretation can be supported by citing further passages and by referring to Johann Schultz, then Kant would have been characterizing arithmetic in three ways by his claim that "arithmetical judgments are synthetic judgments."

1. Arithmetic is an axiomatic science; it requires axioms which can be derived neither merely from the law of contradiction nor from concepts with the help of this law.

2. The results in arithmetic are achieved only by the activity of the mathematician who *calculates* the results.

3. Arithmetic uses symbolic constructions and is thus, in particular, bound up with spatial perception.

3. THE INTERPRETATION OF JOHANN SCHULTZ

It is quite remarkable that Johann Schultz's interpretation has never yet been used in all the voluminous literature about synthetic judgment in Kant, even though Kant himself referred to Schultz as his authentic interpreter and even though Kant read his manuscript and asked for a fundamental alteration.[44] The interpretation goes back to conversations with Kant and sometimes uses the very words from the letters of Kant known to us.[45] I will now examine this interpretation in detail.

In 1784, Johann Schultz published a commentary on Kant's *Critique* with the title *Erläuterungen über des Herrn Professor Kant Critik der reinen Vernunft* (Commentary on Professor Kant's Critique of Pure Reason). This commentary originated as a paper which was expanded into a book at Kant's request. There is a subtitle, *Versuch einiger Winke zur näheren Prüfung derselben* (An Attempt at Some Suggestions for a Closer Examination of It), which originated from a suggestion of Kant's. Schultz's two-volume work, *Prüfung der Kantischen Critik der reinen Vernunft* (Examination of Kant's Critique of Pure Reason), 1789–92, is then this examination itself. Our discussion had just arrived at this point in our own third chapter.

The first volume of the work interprets the introduction, the second, the transcendental aesthetic. The continuation promised never appeared.[46] The first volume, which is our focus of interest here, is organized as follows:

Examination of the Introduction

Section 1. What are judgments a priori? (p. 1)

Section 2. By what signs can we see that a proposition is a judgment a priori? (p. 8)

Section 3. What are synthetic judgments? (p. 28)

Section 4. Are there theoretical sciences which contain synthetic judgments a priori? (p. 45)

I. One part of universal logic is a pure part which consists of pure propositions a priori which are, however, all analytic, not synthetic. (p. 45)

II. Geometry consists of pure, synthetic propositions a priori. (p. 54)

III. The whole of pure arithmetic, and general mathematics in general, is a field of completely pure knowledge, just like geometry, and consists of pure, synthetic propositions a priori. (p. 211)

IV. Physics presupposes a· pure part which is properly called natural science and consists of synthetic propositions a priori. (p. 236)

V. Metaphysics is a pure science consisting of pure, synthetic propositions a priori. (p. 238)

Section 5. The importance of the investigation into how synthetic judgments a priori are possible. (p. 240)

From this classification we can see how strong the interest in geometry is, but afterwards, twenty-five pages are still devoted to arithmetic. The inquiry takes the following course:[47]

Concept and Classification of arithmetic (pp. 211–15); arithmetic is not an empirical science but one a priori (pp. 215–17); everyone holds that propositions of arithmetic are analytic and so a priori (pp. 217–18); Kant was the first to see through the apparently analytic character of arithmetic and to recognize its synthetic character (p. 218).

That arithmetic is synthetic is shown by the following main considerations:

1. Arithmetic has axioms as well as postulates, just like geometry. (pp. 218–19)

2. The whole of arithmetic is based on equations or equalities. (pp. 229–32)

3. Analytic judgments cannot enlarge our concepts at all (p. 232). So the whole of pure mathematics is a synthetic science a priori (p. 233). Mathematics is a science arising from the construction of its concepts (p. 233). Mathematics refers only to objects of the senses (p. 233). Axioms and postulates of time (pp. 234–36).

Schultz therefore maintains that arithmetic is synthetic for three reasons:

1. It is axiomatic.

2. It consists of equalities.

3. It can be expanded.

The question we posed, whether what is essential is in the example of $7 + 5 = 12$ or in that of $7 + 5 = 5 + 7$ is answered in the first two sections to the effect that the synthetic character of arithmetic judgments depends on both factors. Accordingly, Schultz interprets the example twice since he examines the statement the first time from the standpoint of the axiomatic foundation and the second time as an equality.

I have already given an account of this interpretation in detail in chapter 3. It goes as follows: The addition of $7 + 5$ depends on the application of the commutative and associative laws. This squares with the accounts of Hankel, Nelson, and Couturat,[48] although we must not assume that any of these three was acquainted with Schultz's work. Thus Leibniz's attempt to find proofs was shown to be mistaken from Kant's point of view. Shortly thereafter, two attacks on this claim appeared in Johann August Eberhard's *Philosophisches Magazin*. In 1791 Eberhard wrote in an article, "Von dem Einflusse der sinnlichen Anschauungen auf die Wahrheit und Gewissheit":

So the truth of a proposition of arithmetic does not depend on time, i.e., on the order in which the parts of the sum have been thought; 2 and 4 makes 6, and I can think of either the 2 or the 4 first. This follows from the proposition, "Every quantity is equal to all of its real parts taken together," which is based on the fact that the whole is equal to itself and is the same as all of its parts taken together. So how does Herr Schultz bring the perception of time into this proposition, "The sum is equal to its parts"? His first axiom of arithmetic goes as follows: "The sum is equal to its parts, in whatever the order of succession we may think of these parts; 6 is just as much $4 + 2$ as $2 + 4$." I ask anyone whether that does not mean that the truth of a proposition does not depend on time at all, not at all on the order of the time series in which the parts of the sum are thought? The second axiom is similar to the first one, for what it states about the mediate [*mittelbaren*] parts of the sum is only what the first states about the parts in general. It is therefore superfluous and increases the number of axioms of arithmetic unnecessarily.[49]

The second attack comes from Lazarus Bendavid (1762–1832), who himself can be called a Kantian, and who expressly takes this approach at the beginning of his 1791 article "Deduction der mathematischen Prinzipien aus Begriffen":

We can maintain with certainty that arithmetic has no postulates. For the only proposition which refers to mere possibility, namely that thinking about numbers in general cannot be limited, is a formal theorem and can be proved strictly by the principle of identity. . . . But it also has no axioms, for since Freiherr von Wolff proved the theorems usually described as axioms until they were completely evident . . . , there are only two propositions left to prove in order to eliminate all so-called axioms from arithmetic. But in order to understand the proof of the proposition $a + b = b + a$, we must remember that Wolff . . . showed that when equals be taken from equals, equals remain, and . . . if unequals be taken from equals, unequals remain. Assuming this, I maintain: $a + b = b + a$.

For if they were not equal, then $a + b = c$, $b + a = d$, and c and d would be unequal. . . . If a be taken from $a + b$ as well as from c, then $b = e$ results (e means, namely, $c - a$). Likewise, if a be taken from $b + a$ and from d, then $b = f$ where $f = d - a$. Now since c and d are supposed to be different (by hypothesis) and a is taken from both, then e and f should also be different. But since both $= b$, that is impossible.

It follows that $a + b = b + a$. Q.E.D.[50]

Schultz himself dealt with Eberhard's attack in the second volume of his *Prüfung*, which appeared shortly afterward:

But this criticism [of Eberhard's, *Magazin*, IV, p. 69] depends simply on a misunderstanding, for: a) if I think of producing the whole c by adding b to *a*, and on the other hand, of producing the whole *d* when I add *a* to *b*, I obviously do not think of the same things by *c* and *d* but of different wholes, of which the first is the same in concept as *a* + *b* but the other the same as *b* + *a*. It follows from the proposition that the whole is the same as its real parts taken together, that indeed *c* = *a* + *b* and *d* = *b* + *a*, but not that *c* = *d*. This is shown even more clearly in the addition of numbers, for when I add 2 to 4 the sum is derived in a way quite different from the way I add 4 to 2. It is precisely the time sequence, which distinguishes the combining of the parts of the whole *c* from the combining of the parts of the whole *d*, which can very often also make a difference in quality of both, e.g., when a two-cubic-foot pyramid is put on top of a four-cubic-foot pyramid, the whole has a completely different quality from that of a four-cubic-foot pyramid put on top of a two-cubic-foot one. So why need I not fear that there might also be a difference in quantity in them, but instead I can say apodictically that in this respect they are always the same? . . .

Moreover, we should note that it is not always permissible in arithmetic operations to take the parts of the whole one after the other instead of taking the whole all at one time. For I can always say $4/5$ + $3/5$ instead of $7/5$, but I cannot say $7/2$ + $7/3$ instead of $7/(2 + 3)$.[51]

The special significance of the axioms for the synthetic character of arithmetic is again underscored by this discussion.

Incidentally, it is strange that Kant in the *Streitschrift gegen Eberhard* does not deal with arithmetic, and especially its possible axioms. Even though this work does deal mainly with the synthetic character of an a priori natural science, still geometry is examined at length. On the other hand, there is only one extremely tentative remark about arithmetic:

When he considers the objects of inner sense, how does he propose to explain the basic condition of inner sense, time as continuous quantity (like space) but of only one dimension, on the basis of the simple parts of inner sense? According to him, sense perceives the parts but not as *separate* parts, while the intellect thinks of them as separate parts. How does he propose to get such positive knowledge out of the depths of obscurity and so of defects, knowledge which contains the conditions of the sciences which get enlarged a priori more than any other (geometry and general theory of nature)?[52]

This positive knowledge which is the foundation of geometry and of universal natural science is surely pure mathematics. But why is Kant not clearer? Perhaps we can venture the following conjecture: Kant really worked on the *Streitschrift* during the year 1789. Since he may have discussed the synthetic character of arithmetic with Schultz in the same period, between the letter of

November 1788 and the appearance of the *Prüfung* in 1791, Kant might have
avoided positive statements about arithmetic in the *Streitschrift* to allow for the
particular axioms of arithmetic only then becoming clear in these discussions
with Schultz.

Schultz's study is concerned then with the nature of equality in the judgment
$7 + 5 = 12$:

> The whole of arithmetic is based on equations or equalities. Now at first
> glance, an equation, e.g., $7 + 5 = 12$, has very much the appearance not
> only of an analytic proposition but also of an identity in which the subject
> and predicate are one and the same concept, or even that the one is the true
> name of the other. Hence people usually explain the equality $7 + 5 = 12$
> thus: $7 + 5$ *is* 12, or even, as H. Reimarus would have it, $7 + 5$ is *called*
> 12.[53] But if this were correct then it would not be necessary to add at all, for
> anyone who simply knew the number would have to think of the concept of
> the number 12 immediately in the concept of $7 + 5$, so that it would be as
> impossible not to think of the number 12 when thinking of $7 + 5$ as it is for
> us to think of the completely perfect being without thinking of God. . . .
> Furthermore, since the number 12 can also be the predicate in innumerably
> many other equalities, e.g., $18 - 6 = 12$, $3 \times 4 = 12$, $144 = 12$, etc., it is
> also the case that $7 + 5 = 18 - 6 = 3 \times 4 = 144$, etc. So $7 + 5$ and $18 -
> 6$, 3×4, 144, and innumerable other combinations of the same sort would
> *only* be *one* concept, and I would have to think of all of these as well
> whenever I think of $7 + 5$. . . . Thus it is evident that the mere analysis of
> the concept of $7 + 5$ can never lead me to the conclusion that its quantity is
> as much as that of 12; but if I am to recognize this, I simply have to go
> beyond this concept and first use the two axioms of arithmetic for help,
> together with the first postulate. These axioms teach me that I am free to
> add the parts of 5 to 7 successively instead of adding it all at once as a
> whole, and thereby at the same time changing the order of the parts. The
> postulate shows me that and how this is possible. So I first come to the
> insight that the quantity of $7 + 5$ is the same as that of 12 through this
> laborious synthetic procedure.[54]

One needs to observe here that Schultz speaks of a "laborious synthetic
procedure" and insists that this "synthetic procedure" must mean something
different than the synthetic character of the axioms. Until now we have seen that
arithmetic is synthetic because based on synthetic axioms, and we will see later
that the axioms are synthetic because they are based on perception, the axioms
of arithmetic on the perception of time. However, according to the present
interpretation, a "laborious synthetic procedure" refers to the fact that arithmetic depends on an actual calculation by specific procedures. Let us take
another example; the proposition $(a + b)^2 = a^2 + 2ab + b^2$ is derived in the
following way:

$$(a + b)^2 = (a + b)^2$$
$$= (a + b)(a + b)$$
$$= a(a + b) + b(a + b)$$
$$= a \times a + a \times b + b \times a + b \times b$$
$$= a \times a + a \times b + a \times b + b \times b$$
$$= a^2 + 2ab + b^2$$

We see that this is actually the fundamental character of the arithmetic process: an equation of identity is set up and then the right side—or else both sides—can be converted according to the axioms or laws available, until the desired equality is obtained. It would be easy to use a somewhat more complicated example, say a proposition of integral calculus. The basic process is shown with special clarity when a symbolic representation is used exclusively, as in the *Principia Mathematica* of Russell and Whitehead.

The argument given here for the synthetic character of arithmetic, namely, that the procedure is synthetic, corresponds exactly to Kant's own line of thinking as he describes it to Schultz in the 1788 letter cited earlier:

> I can form concepts for myself in various ways from the very same quantities by combining and separating. (But both addition and subtraction are syntheses.) These concepts are, indeed, objectively identical (as in every equation) but subjectively, according to the kind of combination I am thinking of in order to obtain that concept, they may be very different; so that the judgment goes beyond the concept which I have from synthesis since it substitutes another kind of synthesis (which is simpler and more appropriate for the construction) in the place of the first one, a synthesis which still defines the object in the very same way. In this way I can determine one and the same quantity $= 8$ as, $3 + 5$, as $12 - 4$, as 2×4, as 2^3. But the thought of 2×4 was not contained at all in my thought of $3 + 5$, but then neither was the concept of 8, which has one and the same value as both, contained in it.[55]

Schultz gives a third reason for the view that arithmetic is synthetic, namely, that arithmetic is a science which is expanding enormously. The section in Schultz about this corresponds practically word for word with the apposite paragraph in Kant's letter. I have already cited the passage in chapter 3 (p. 38) because of the connection between Kant, Schultz, and Murhard.

Another consideration can be added to the last two arguments, one which I at least want to indicate here. If actual calculation plays such a role in arithmetic and if arithmetic is thus a science capable of great expansion, the question immediately arises of the significance of this activity of the mathematician for arithmetic as a whole, whether there is any arithmetic at all before such activity, or, as the common saying is, whether propositions of arithmetic are discovered or invented. Schultz's answer is unambiguous:

> Now the true nature of the *necessary* and *eternal* truths of arithmetic also manifests itself. Each proposition of arithmetic is *objectively* true, so that its truth does not depend on the particular state of the subject who is thinking of it, but instead every subject who understands it must positively hold it to be true. Number is not a thing which exists in itself but is an object which first has to be produced by the successive combining of units by the intellect . . . and every proposition of arithmetic only arises in this way. . . . So it is obviously contradictory to believe that, e.g., since the proposition 2 times 4 is 8 is objectively true, the product 8 of 4 and 2 is not only produced or brought forth by multiplication but has also been so eternally; for this would be to say that the number 4 has already really [*wirklich*] been multiplied by 2 and has yielded the product 8 even before it has been multiplied by 2.[56]

To conclude this matter, we once more want to go through the train of thought in the long letter of Kant to Schultz. We remember that Schultz had sent the manuscript of the *Prüfung* to Kant and that the judgments of arithmetic had been described in this manuscript as analytic. In the letter of November 1788, Kant urges Schultz to re-examine his point of view more thoroughly. There are five parts to Kant's argument that arithmetic is synthetic:

1. General arithmetic is a science which is expanding. (See above, p. 38.)

2. Different representations of any quantity are possible. These are, indeed, objctively identical but are subjectively different because they are different kinds of combinations. Thus 8 can be represented by $3 + 5$, by $12 - 4$, 2×4, by 2^3. (See above, p. 105.)

3a. The judgment $3 + 4 = 7$ is a postulate.

3b. If I have to think of 7 at the same time as I think of $3 + 4$, then I also have to think of $12 - 5$; but this is not consistent with what I am conscious of doing.

4. Thus the meaning of an equation also consists in representing the same objective quantity in different ways, in constructing it in different ways.

5. Arithmetic and time.

We see then that Schultz's second argument from procedure and his third, from the fact that our knowledge of arithmetic can increase, are closely connected to Kant's, sometimes even verbally. As to the first argument derived from the axiomatic character of arithmetic, there is agreement about the postulates, since Kant in his letter also argues that arithmetic is synthetic because of the postulates. On the other hand, from the material known to us nothing certain can be asserted about the two axioms—the commutative and the associative—as I have already indicated.

C. The History of the Interpretations of Arithmetical Judgment
I. KANT'S CONTEMPORARIES

A lively discussion about synthetic judgment developed even in followers of Kant, who worked out the fundamental problems clearly and carefully, even though these explanations were mostly just a paraphrase of Kantian remarks. Probably one of the first investigations was the work of Christian Gottfried Schütz (1747–1832), *Programma de syntheticis mathematicorum pronuntiationibus*, which appeared in 1785:

> Likewise in arithmetic all arguments are developed by constructing and separating signs and characters [*characterum*], and in the same way, everything in this part of mathematics which is obvious to everybody also goes back to perception (*ad intuitum*).[57] There is a theorem: In a geometric progression the product of the first and last terms is equal to the individual products of a pair of terms equally distant from the first and the last in each direction. At this point, if some philosopher were to begin to occupy himself early in the morning with explaining the subject and adequately revealing the force of it in words, it would be bedtime at night before he would get to the predicate, as that character in Plautus remarks.[58] What of the mathematician? After he has named the first term with the general name a, and the exponent of the ratio or the index, he then examines the series with the same subscripts [i.e., superscripts] in inverse order:
>
> $$a \; am \; am^2 \; am^3 \; am^4 \; am^5 \; am^6$$
>
> $$am^6 \; am^5 \; am^4 \; am^3 \; am^2 \; am^1 \; am^0$$
>
> Hence it should be clear to him as he observes that the product of the first and seventh or the last is equal to the product of the second and the sixth, in the same way of the third and the fifth, and finally of the fourth times itself—clearly $a^2 \; m^6$.[59]

Schütz then draws attention to the special importance of the primitive perception of fingers and points.

Andreas Metz (1767–1839) in a work of 1795, *Kurze und deutliche Darstellung des Kantischen Systems*, also goes in the same direction:

> I cannot discover that $108 + 2389 = 2497$ or that $2389 - 108 = 2281$ by analyzing the concepts of $108 + 2389$ or of $2389 - 108$ through all eternity. . . . I do not know what the sum or difference is before I go outside these concepts and have recourse to perception, namely, that I add the units of the first to the units of the other one after the other, on the leading strings of perception (say, by points).[60]

The works of Georg Samuel Albert Mellin (1755–1825) had a marked influence at this time. The most important is probably the six-volume *Enzyklo-*

pädisches Wörterbuch der kritischen Philosophie (1797–1804). Although
Mellin knew of the account of the axioms through Schultz and put it in the article
"Axiom," he makes no use of the basing of synthetic judgment on the axiomatic
point of view. Instead, he relies solely on the symbolic construction in the
detailed discussion of arithmetical judgment.[61]

Mellin then offers a second justification:

> The two concepts (of 7 + 5 and of 12) are connected to each other for they
> are just like Augustus and Caesar, concepts of the same object, but they do
> not present it to thought by the same characteristics. This notion about
> different characteristics leads to the necessary consequence of making the
> difference subjective.[62]

In spite of these explanations by his followers, Kant's views about the
synthetic nature of arithmetic judgment met with strong opposition. I have
already referred to the attacks of Eberhard and Bendavid (see above, pp. 101–2).
Johann Georg Heinrich Feder (1740–1821) dismissed the distinction [between
different characteristics of the same object] from the very first as not worth
discussing with the rhetorical question: "Isn't it [the proposition that 7 + 5 =
12] obviously analytic?"[63] Dietrich Tiedemann (1748–1803) was one of the
first to try to refute Kant's views in his 1785 *Über die Natur der Metaphysik, zur
Prüfung von Herrn Professor Kants Grundsätzen:*

> Consequently, the proposition that 7 + 5 makes 12 will have to be analytic.
> 7 and 5 are also to be found among all the possible parts of 12, so that to say
> that 7 and 5 are 12, it is unnecessary to go outside the concept of 12.[64]

Tiedemann here obviously confuses the parts of the concept with the parts of
the thing; 7 and 5 are indeed parts of 12, but they are not thereby parts of the
concept of 12. This becomes clearer in geometry. One part of a square is, say, a
half a square; a part of the *concept* of square is—according to one definition,
"equal-sided rectangle"—equal-sided. We have already seen that and how,
according to Leibniz, these two *parts* can coincide in one specific theory of
judgment. It does not work, however, the way Tiedemann tries to do it.

Johann Christoph Schwab took part in the prize competition of the Berlin
Academy for which Kant wrote drafts of the *Advances In Metaphysics*.
Schwab's prize-winning work, published in 1796, contains an appendix "Von
den analytischen und synthetischen Urteilen" (On Analytic and Synthetic
Judgments):

> The proposition 1 + 1 = 2 is supposed to be synthetic and therefore true
> because I add the unit successively to itself so that this operation takes

place in time. This adding together may take place successively and in time in our mind, but this succession is not the reason that the proposition is true. The proposition $1 + 1 = 2$ is obviously an identity, and it needs nothing more to satisfy us of its truth. It ultimately follows from this Kantian theory of the foundations of arithmetical truths that there is no arithmetic for a mind which is not restricted to the conditions of time in its ideas. Until now all philosophers have admitted that the divine mind does not count as we do, i.e., that it does not formulate and think of the numbers and truths of arithmetic successively. But that there are no numbers at all and no arithmetic for the divine mind—such a position was reserved for the new philosophy.[65]

In this attack, Schwab has really gotten to the heart of the Kantian line of thought. If Kant traces the essence of the human mind to its finiteness, then all of mathematics becomes really a concern of man and only a concern of this finite mind.

Gottlob Ernst Schulze (1761–1833), well-known at the time, published a two-volume criticism of Kantian theory in 1801. In this work, *Kritik der theoretischen Philosophie*, he argues at the outset that the axioms of arithmetic given by Johann Schultz are not unprovable and then tries to prove these propositions from the mere concept of a number.[66] In the second volume the attack is newly formulated in an interesting way:

Concepts are only distinguished from one another by their content, not by their signs, spoken or written. Concepts of numbers (in which numbers are abstracted from all quality of the units combined in them) must therefore be distinguished either by the set of units combined therein or with respect to the combination of these units. There are, however, no more units thought of in the number 12 than in the concept of $7 + 5$, nor are they thought of as differently combined.[67]

In this last point Schulze errs. It is, of course, correct that no more units are thought of in 12 than in $7 + 5$. But these units, contrary to his claim, are combined in the two forms in different ways: in 12 as $1 + 1 + 1 + 1 + 1 + 1 + 1 + 1 + 1 + 1 + 1 + 1$ as opposed to $7 + 5$ as $(1 + 1 + 1 + 1 + 1 + 1 + 1) + (1 + 1 + 1 + 1 + 1)$. There can be no doubt that these two kinds of combination are different. Johann Schultz arrived at precisely this solution, for in the *Prüfung* he analyzes $7 + 5 = 12$ as $7 + (1 + 1 + 1 + 1 + 1)$, breaking just one of the parts into units, and as he correctly states, we can only go further by using the axioms.[68]

2. THE NINETEENTH CENTURY

We must always guard against sliding into purely logical distinctions in discussing analytic and sythetic judgment. The difference between the two is deprived of all meaning as soon as we lose sight of the ontological character of arithmetic—which is the part of the problem we are dealing with. But this is the wrong turn most scholars have taken. The discussions of Kant's account of analytic and synthetic judgment are divided into seven groups:

1. German Idealism

2. Bernhard Bolzano and Edmund Husserl

3. The mathematicians

4. The difference is interpreted as relative: Friedrich Schleiermacher (1768–1834) and practically all the logicians and historians of philosophy.

5. Close adherence to Kant: Aloys (Alois) Riehl (1844–1924)

6. All propositions become synthetic, even the analytic principles.

7. Perception is rejected: the Marburg school.

I shall content myself with a brief survey of groups 1, 4, 5, and 6 since an exhaustive account would go on forever.

The philosophical objective of German Idealism brings practically nothing to bear on the question of the synthetic character of arithmetic. First, any positive relation to mathematics has gotten lost; then, when mathematical questions are discussed at all, the more accessible geometry is chosen. What is decisive, however, is that the basic approach of a closed deductive system is incompatible with the axiomatic character of mathematics as Kant conceives it. If mathematics has axioms, then it depends on numerous, underivable principles and, if so, it cannot be deduced in any way. But Fichte insists on a deduction not only of mathematics but of all nature. Thus he says in the first edition (1794) of *Über den Begriff der Wissenschaftslehre*:

> [The laws of nature] can be derived from the basic principles of human knowledge prior to any observation, the least just like the greatest, the construction of the most insignificant blade of grass just like the movements of the heavenly bodies.[69]

It is really absolutely impossible to make this view of Fichte's compatible with the position Kant had so carefully formulated.

It is not much better with Hegel. Even if Hegel's 1801 habilitation *De orbitis planetarum* (On the Orbits of the Planets) is not already a stumbling block for anyone with a scientific background, the numerous definitions and observations in the 1817 *Enzyklopädie* are totally unacceptable. It must be admitted that Gauss is correct when he writes in a letter to Heinrich Christian Schumacher (1780–1850): "Look around you at the philosophers of today, Schelling, Hegel, Nees von Esenbeck, and company. Don't their definitions make your hair stand on end?"[70] This question seems important to me because the roots of the deep and disastrous estrangement between mathematics and philosophy lie here. The rift was not created by Kant but by the German Idealists. Here we will not think so much about the line of argument by which Hegel pronounces mathematical concepts "external" or "arbitrary" [*gleichgültige*] as in the *Wissenschaft der Logik*: "For number is arbitrary determinateness. . . . " No one will want to deny the core of truth in this pronouncement, but Hegel lacks any real interest in or understanding of mathematical problems. The extensive chapters in the *Logik* about the calculus merely present the work of the French mathematicians in an unoriginal way. While Kant avoided the discussion of the infinitely small, Hegel lets himself be dragged into the discussion about the infinitesimal, which at that time was still purely mathematical. Accordingly, Hegel's attack on synthetic judgment in arithmetic stands on weak ground and, to my mind, cannot even be compared in intellectual depth with the attacks of Eberhard, Schwab, or G. E. Schulze.[71]

Hegel writes in the *Wissenschaft der Logik*:

The concept of the sum signifies no more than the abstract determination that these two numbers *ought* to be combined, and that is as numbers in an external way, i.e., without concepts—that we are to count from seven on until the ones added on come to five, the result leading to the name otherwise known as twelve. . . . Five, it is true, is given in perception, i.e., in an entirely agglomeration [*Zusammengefügtsein*] of the thought of one repeated arbitrarily; but neither is seven a concept; there are no concepts beyond which a transition is made. The sum of 5 and 7 means only the combining of the two numbers *without* concepts, the continual enumeration *without* concepts from seven until the five are used up; we can call combining, like the enumeration from one on, synthesizing—but a synthesizing that is totally analytic in nature since the connection is wholly artifical, nothing is in it nor enters it which is not at hand quite externally. The postulate [*Postulat*] of adding 5 to 7 is related to the postulate of enumeration in general, like the postulate of producing a straight line is to that of drawing a straight line.[72]

It is curious that even the works of Jakob Fries are not very productive of a further explanation of synthetic judgments, although he has a strong interest in mathematics. There are apparently two reasons for this, one being that all too often, as in the *Neue Kritik der Vernunft* of 1807, Fries lets himself be diverted into the logical description of the distinction and so completely misses the basic significance of Kantian perception [*Anschauung*].[73] Later he devotes himself almost exclusively to combinatorial matters. Combinatorics, indeed, even occupies the greater part of his 1882 *Die mathematische Naturphilosophie*.

The fourth group of interpretations has its origin in Friedrich Schleiermacher. From the point of view of his *Dialektik* of 1839 (from section 308 on), the difference between analytic and synthetic is only subjective and relative:

The difference between analytic and synthetic judgments is not hard and fast and does not come into question for us at all.

One and the same judgment (ice melts) can be analytic if the origin and disappearance at certain temperatures are already assumed in the concept of ice, and it can be synthetic if not. This, however, also holds true for complete judgments since a particular group of things taken together can also be brought under a concept, e.g., every world system is such a concept. So this difference only expresses a different state of concept formation.

Note: This is also valid for mathematical judgments. The proposition about the angles of a triangle is analytic only if the concept of a triangle includes its origin from the motion of a line down from the vertex of an angle, although the other simple motions have to precede it.[74]

Practically all the researchers of the nineteenth century followed Schleiermacher. I mention here only in passing Friedrich Albert Lange (1828–1875), Christoph Sigwart (1830–1904), Friedrich Adolf Trendelenburg (1802–1872), Friedrich Überweg (1826–1871), Benno Erdmann (1851–1921), Ernst von Aster (1880–1948), and Eduard von Hartmann (1842–1906).

Then there is a fifth group of interpretations which is so dependent on Kant that nothing new results. Aloys Riehl is an example. The explanations of synthetic judgment to be found in the first volume of his *Der philosophische Kritizismus* are virtually but a paraphrase of the Kantian propositions.[75]

A sixth group takes the concept of synthetic judgment so broadly that in the end all propositions, and the Kantian analytic principles in particular, become synthetic. The approach that goes furthest in this direction is probably that of Wilhelm Wundt (1832–1902), who deals with the axioms in this regard in the fourth part of the first volume of his *Logik of 1880–83*. A more detailed explanation is to be found in a dissertation (inspired by Wundt) of 1880 by Willi Reichardt (1864–1924), *Kants Lehre von den synthetischen Urteilen* (Kant's Theory of Synthetic Judgment).[76] Gottlob Lipps (1865–1931) has a similar interpretation in his *Untersuchung über die Grundlagen der Mathematik*. Heinrich Rickert (1863–1936) also belongs here in a sense. When Rickert tenaciously maintains in his dispute with the Marburg school that there is a nonlogical factor in mathematics, he may be much closer to the literal meaning of Kantian perception [*Anschauung*] than the people from Marburg. On the other hand, when he states that the proposition $1 = 1$ is synthetic, he certainly goes beyond Kant.[77]

3. THE MATHEMATICIANS

Gottlob Frege gives a detailed explanation of arithmetical judgment in his *Die Grundlagen der Arithmetik* (The Principles of Arithmetic). First of all, he deals with Leibniz's proof:

This proof [by Leibniz] at first seems to be constructed wholly from the definitions and axioms introduced. . . . With more careful inspection, a gap is found which is hidden by omitting the parentheses. It has to be written more precisely:

$$2 + 2 = 2 + (1 + 1)$$
$$(2 + 1) + 1 = 3 + 1 = 4$$

Here the following proposition is missing:

$$2 + (1 + 1) = (2 + 1) + 1$$

which is a particular case of

$$a + (b + c) = (a + b) + c$$

If this rule is presupposed, we can easily see that every formula of the one-plus-one type can be proved in this way.[78]

This gap in Leibniz's proof is also made clear by Leonard Nelson in his lecture on "Kant und die nichteuklidische Geometrie" (Kant and Non-Euclidean Geometry).[79] Frege also notes that Kant, as we have already seen, (see above, p. 91), holds that the numbers are not homogeneous but distinguishes between small and large numbers:

Kant obviously had only small numbers in mind. The formulas for large numbers would then be provable, while they are immediately obvious for the small ones. But in principle it is hard to make a distinction between small and large numbers, especially since no sharp boundary can be made. If the number formulas from, say, 10 on were provable, it could properly be asked, why not from 5 on, from 2 on, from 1 on?[80]

Frege has not really grasped the exact meaning of the Kantian claim, for Kant never says anywhere that the addition of small numbers is immediately obvious. Moreover, Frege has let himself be forced into the same misunderstanding of the passage at B15ff. we have already seen in Hankel (see above, p. 98):

Kant wants to use the perception of fingers or points as an aid, and here he falls into the danger of letting these propositions appear to be empirical, contrary to his intent; for the perception of 37,863 fingers is, in any case, not a pure one. Neither does the expression perception seem to fit. . . . Do we have any perception at all of 135,664 fingers or points? If we did, and we had one of 37,863 fingers and one of 173,527 fingers, it would have to be immediately obvious that our equality was correct [134,664 + 37,863 = 173,527], at least for the fingers, if it were unprovable; but this is not the case.[81]

Frege did, indeed, express very high regard for Kant's mathematical work, but these misunderstandings lead him to hold that only geometry is synthetic, while he takes arithmetic to be analytic, as against Kant.[82]

Louis Couturat follows Frege closely in his 1904 article "La philosophie des mathématiques de Kant." He, too, comes to the conclusion that Leibniz's proof has a gap but maintains in spite of this that all judgments of arithmetic are analytic.

I do not know whether Paul Mansion at least recognized the significance of the two axioms for the addition of integers in his 1908 "Gauss contre Kant sur la géometrie non euclidenne": "In the *Critique of Pure Reason*, because of his ignorance of the definition of the sign $+$ and of 5, he does not see that $7 + 5 = 12$ can be proved by a series of analytic judgments."[83]

Otto Hölder (1859–1937) in his 1924 *Die mathematische Methode* rejects the axiomatic method and advocates a genetic method whose philosophical position is not easy to determine. According to him, there are axioms only in geometry. He thus derives the commutative and associative laws of addition from definitions, like Grassmann. But he thereby gets into the difficulty of interpreting the continuum, and so a concept of arithmetic, either as indeed axiomatic or "as a given original form." From here on he comes to grips with Kant.[84]

A discussion of David Hilbert, L. E. J. Brouwer, and Bertrand Russell is pointless so long as the geometrical questions cannot be referred to. The basic features are clear only in Russell. *Principia Mathematica* begins with ten "primitive propositions"; 1.4 introduces the commutative law and 1.5 the associative; 1.6 is at bottom the Modus [syllogism] Barbara and arises from this principle by means of some brief further development. Using this assumption, Russell proves the law of contradiction, formulated as the law of excluded middle. It is possible to interchange certain principles and certain theorems and also possible to condense these principles into a single proposition most ingeniously. Whether such a condensation (which, incidentally, does not stem from Russell himself) is satisfactory is disputable.

The question is how *Principia Mathematica* relates to Kant's idea of the basic nature of mathematics. Here the answer is that this treatise is the purest realization of the Kantian idea of pure mathematics. There are four decisive reasons for such a judgment. *Principia Mathematica* eliminates all reference to particular quantities; it is axiomatically based; it uses a purely symbolic construction; and it is a science capable of expansion.

I cannot agree, therefore, with the opposite conclusion of Heinrich Behmann (1891–1970) that in *Principia Mathematica* the analytic nature of mathematics is conclusively proved. We should not be drawn into a dispute about words here.

If logistic did emerge as a new science, above all the question arises about the position it must occupy in the Kantian classification of the sciences. If Kant expressly asserts that formal logic uses only the principle of contradiction, then it is impossible for logistic to be considered formal logic in the Kantian sense, insofar as logistic uses the associative and commutative laws. From the very beginning, therefore, I rule out the assumption that since mathematics is a part of logistic, it can or must be regarded as part of formal logic. But what Behmann maintains pertinent to this is also inconclusive. From the outset he lets himself be dragged in the wrong direction by defining the difference between analytic and synthetic judgments in a purely logical manner. In his essay "Sind die mathematischen Urteile analytisch oder synthetisch?" (Are Mathematical Judgments Analytic or Synthetic?) of 1934 he says:

> [I]t appears to me that Kant's real intention has been correctly described in what I have said heretofore, and that in order to see this clearly, it is only necessary to get rid of certain of the awkward traits and obscurities which plague his exposition.[85]

For my part, I emphatically deny the need to get rid of obscurity or awkwardness in Kant's exposition. Quite the contrary, no philosophical work exists which is so cleanly worked out as the *Critique of Pure Reason*.

Behmann starts from three fundamental ideas. He maintains, first, that Russell presents mathematical problems in a purely logistic way:

> Because of the reasons already given we will have to understand by an analytic judgment one which is presented as a purely logical law or as an application of one after the definitions in terms of basic concepts have been substituted for the given defined concepts.[86]

If Behmann wants to say here that there are complex concepts and simple, purely logical, basic concepts, he will certainly not find any support for this in Russell himself. But if he wants to refer to the most conspicous characteristic of the *Principia Mathematica*, namely, that mathematical problems are handled there in a purely logistic, i.e., a purely symbolic language of signs, then this observation in no way tells against Kant. Kant repeatedly said that symbolic construction was the mark of pure mathematics. And no one can doubt that the *Principia Mathematica* is the best form of a purely symbolic construction up to the present day.

Behmann finds a second reason for the analytic character of mathematics: that arithmetic can only make use of logical principles.[87] According to Kant, there is no plurality of principles in an analytic logic at all; by his express statement, analytic logic has only one single principle, namely, the principle of contradiction. All other analytic propositions must be proved, as Kant indicated for analytic principles (see above, chap. 2). When logistic takes the commutative and associative laws as among its principles, it introduces synthetic axioms in Kant's sense and emulates him.

After these two inadequate reasons, a third might be successful: "Namely, a new construction of formal logic is required, not as a more or less clear arrangement but as a closed and clearly surveyed system of concepts and laws."[88] If it were really the case that logistic, and with it modern mathematics, were a system of concepts and laws, then a conclusive argument against Kant's interpretation of mathematics would have been found. If mathematics were a system, on the one hand, the fundamental propositions would have to be brought into a systematic relation, which would mean either that some of them would be subordinate to others or that they would be derivable from one universal principle; on the other hand, through such a system mathematics would lose its character as a science which expands enormously. A system, indeed, means a complete and closed synopsis of a field; nothing basically new can further arise in a system. If the *Principia Mathematica* be examined, it becomes in no way a closed and clearly surveyed system, as Behmann thinks. The fact that Russell has to introduce 300 basic symbols alone is evidence against this. A systematic relationhip is just as little to be seen between the principles, not to speak of saying that Russell saw any such thing. Obviously, Behmann arrived at incorrect results by letting himself be drawn into defining the difference in terms of logic. The three basic characteristics by which Kant interpreted mathematics—axiomatics, symbolic construction, and expansion— are not refuted by logistic but confirmed by it.

4. BOLZANO AND HUSSERL

I have already explained how Bernhard Bolzano defines ideas-in-themselves and truths-in-themselves (see above, p. 87). For further details I cite a passage from the 1850 *Neuer Anti-Kant* of Frantisek Prihonsky (1788–1859):

> If we speak in this way, obviously by *true* or *truth* we are thinking of a particular proposition-in-itself, regardless of whether there is anyone who holds this proposition to be true and says so, even if he thinks of it only to himself, or even if there isn't any such thing. We thereupon take the word *truth* in a sense according to which there are truths-in-themselves, and so also propositions-in-themselves, which no one (except God) knows or even thinks of. So we say, e.g., that before anyone has even asked the question of what kind of digit is in the thousandth decimal position of the number π, it is, nevertheless, one of the ten propositions we think of when we explain that it is true or is a truth that the numeral referred to is a zero, a one, a two, etc.[89]

God eternally thinks of all these mathematical propositions.

> From the omniscience of God it follows, indeed, that every truth is known
> to the Omniscient, even if not known by any other being, even thought of,
> and it is perpetually present to His Mind. Therefore, there is really not a
> single truth which is not known by anyone at all.[90]

I have already explained that the two contrary interpretations cannot be recon-
ciled, namely, between this interpretation of numbers as ideas-in-themselves
eternally thought of by God, and the Kantian interpretation of number as
quantitas phaenomenon.

After Bolzano, starting from the proposition-in-itself, develops the concept of
an idea and an idea-in-itself,[91] he distinguishes ideas in two respects. In one of
them, ideas are either simple or complex; in another, they have either one or
more than one object. Bolzano calls ideas which are simple and have only one
object *perceptions* [*Anschauungen*];[92] ideas which are not perceptions and also
contain no parts which are perceptions are called *concepts* [*Begriffe*];[93] complex ideas which contain some perceptions as parts are called *mixed ideas*
[*gemischte Vorstellungen*].[94] Bolzano believed that he was carrying on what
Kant had begun; and a certain relation to Kant's definition of perception as
repraesentio singularis is certainly apparent. But the results cannot be recon-
ciled with each other since Bolzano, because of his definition, starts with space
and time as pure concepts: "But if the ideas of the whole of infinite time and of
the whole of infinite space are not perceptions, then they are pure concepts."[95]

The account of the difference between analytic and synthetic judgments is
also developed in an entirely different way:

> But if there is only a single idea in a proposition which can be changed
> arbitrarily without affecting the truth or falsity of the proposition; that is, if
> all propositions which arise from interchanging their ideas with others
> arbitrarily are either altogether true or altogether false, provided only that
> they have objectivity: this characteristic of the proposition is by itself
> unusual enough to distinguish it from all those for which this is not the
> case. So I allow myself to identify propositions of this kind with a term
> borrowed from Kant, *analytic*, but all the remaining, that is, those for
> which there is no single idea that can be changed arbitrarily without
> changing their truth or falsity, I call *synthetic*.[96]

Oddly enough, in making this distinction, Bolzano sets the highest value on
the relationship to Kant:

> Indeed, one who went most deeply into this distinction we are speaking of,
> whom the author of this book must thank if his own views on this subject

should be correct, is Kant. . . . He maintained that every area of knowledge dealt only with knowledge of these synthetic truths; that all the theorems of mathematics, physics, etc. were such synthetic truths. Whoever recognizes this as correct is also close to seeing that there are innumerably many characteristics of an object which can be derived apodictically from its concept although we do not think of them as parts of this concept at all.[97]

Despite this weighty statement, I still believe that in the final analysis there is only a terminological agreement. The importance of perception in mathematics is so essential for Kant that to abolish perception means also to abolish agreement with Kant. In fact, like others, Bolzano interprets perception for Kant as being appearance:

> From these and similar examples I take it that Kant attributed to the effect of a special, pure perception what is assumed from certain, only obscurely given grounds, mostly from the mere evidence of the senses, particularly of the eyes.[98]

Such an interpretation of Kantian perception was, of course, easy for Bolzano, whose own mathematical work succeeded in supplying a purely arithmetic exposition of the basic concepts. I hope I have already sufficiently shown, however, that Kant already knew that this work had to be done. The basic difference between Bolzano and Kant is over the truths-in-themselves. If mathematics is such a realm of truths-in-themselves, then not only does the fundamental significance of the axioms become blurred but also the significance of the arithmetic method as symbolic construction diminishes, since in that event the mathematical truths are no longer constructed but are taken as truths-in-themselves. In his remarks about the arithmetic judgment $7 + 5 = 12$ itself, Bolzano wants to define the concept of the sum by means of the associative and commutative laws and then to derive the individual sums analytically.[99]

A whole series of researchers was under the influence of Bolzano. There is first Prihonsky, whose *Neuer Anti-Kant* is an explanatory epitome of the *Wissenschaftslehre*. Prihonsky examines the whole of the *Critique of Pure Reason* in relation to the *Wissenschaftslehre*, compares it section by section with Bolzano's views, and pays particular attention to mathematical problems. The contrast between Kant and Bolzano appears rather sharper in Prihonsky's account than would be expected from what Bolzano himself says.

The 1871 "Über Kants mathematisches Vorurtheil und dessen Folgen" (On Kant's Mathematical Prejudice and Its Consequences) by Robert Zimmermann (1824–1898) also seems to have had a great influence; it is a work which Husserl still applauds in his 1891 *Philosophie der Arithmetik*. The following sentence illustrates Zimmermann's approach: "The view that all mathematical judgments are synthetic can be called Kant's mathematical prejudice just as Fries, as is well known, spoke of his transcendental prejudice."[100] Zimmermann's objections to Kant are utterly pathetic, and one could even think that Zimmermann had never recognized the significance of the associative and commutative laws for the proposition $7 + 5 = 12$. Such an omission would, of course, have been unforgivable even at that time. Julius Bergmann (1840–1904) ought also to be mentioned here. In his work "Über den Satz des zureichendes Grundes" (On the Principle of Sufficient Reason) of 1900, he discusses arithmetic judgment in detail. There he uses Leibniz's proof unperturbed, without noticing the missing associative law.[101]

Franz Brentano (1838–1917) gives detailed explanations of mathematics. The second part of the *Versuch über die Erkenntnis* (Essay on Knowledge) of 1925 deals with the logical character of arithmetic. According to Brentano, mathematics starts first from empirical concepts. For him, clear judgments are divided into two classes, one being that of perceptions of obvious, indubitable facts. "The others are universal negative judgments which reject the object of certain complex concepts as impossible."[102] These universal negative judgments fall into three classes of the following form:

> It is impossible for something red to be blue.
> It is impossible for something blue not to be blue.
> There is no bridegroom without a bride.

According to Brentano, all mathematical propositions are of the form "There is no *C* which is *A* and *B* at the same time." In this sense, mathematics—geometry as well as arithmetic—is a purely logical science. The theorems as well as the principles are analytic propositions carrying apodictic evidence, because they are all merely cases of the law of contradiction. For Brentano, mathematics is not thereby resolved into a system of conditional propositions of the form "If such and such axioms are valid, then such and such theorems are valid." Rather, the axioms are absolutely certain as analytic truths and can be proved. There is only one real and absolute geometry, the Euclidean.

In this way Brentano arrives at an absolute mathematics in spite of his starting with the empirical. Analytic propositions which are apodictically evident can be extracted as axioms from particular, empirically given concepts, and the whole of mathematics from these axioms purely syllogistically. For Brentano, then, Kant's interpretation of mathematics is obviously nonsense.

Edmund Husserl develops Bolzano's approach to perfection and so completes the return to Leibniz. The basic concept is the concept of a clearly defined plurality, with which I began in the introduction to this work. Accordingly, when Husserl interprets mathematics, be it in part or as a whole, as a clearly defined plurality, he means that mathematics is a system of propositions free from contradiction, and derived from a specific number of principles purely syllogistically. Starting from this, Husserl gives a detailed exposition of mathematical problems in the *Formalen und transzendentalen Logik* (Formal and Transcendental Logic) which appeared in 1929. There he insists on Leibniz's *mathesis universalis* in the sense of a purely analytic kind of knowledge:

> The new highest concept of the discipline here in question would therefore be the form of a deductive theory or a "deductive system." . . . Besides the task of its formal definition, there is the endless task of sketching the possible forms of such theories, of differentiating the forms subsumed under it in an explicit, systematic arrangement, and also of seeing the various forms of theories of this sort theoretically as instances of higher universalities of form, of differentiating in a systematic theory the particular determinate forms subsumed under each of these higher universalities of form and at the very pinnacle, indeed under the highest idea of any form of theory in general, a deductive system in general.[103]

> The idea of a universal task emerges here: to strive toward a highest theory which includes all possible forms of theory or all possible forms of plurality as mathematical particulars and so deducible.[104]

> At first when we ask about the concept of the universal which should circumscribe the homogeneous area of these obviously related disciplines, we are perplexed. But if we consider the naturally widest universality of the concepts of set and number and the concepts of element and unity which respectively determine their meaning, we recognize that the theory of sets (*Mengenlehre*) and numbers refers to the empty universe, object in general or something in general. . . . This gives rise to the idea of a kind of universal science, that of a formal mathematics in the all-inclusive sense, whose universal field is firmly delimited as the field of the object of the highest concept of Form-in-General or the Something-in-General of the emptiest generality, with all forms of derivation producible in the field a priori and therefore thinkable. . . . Such derivations are in addition to sets and numbers (finite and infinite), combination, relation, series, union, whole and part, etc. So as an ontology the whole of mathematics is close at hand . . . but as a formal one, to be regarded as referring to the pure mode of the Something-in-General.[105]

Husserl thus insists on a *mathesis universalis* as a system of all possible systems; this is a determinate plurality and deals with the possible variations of Something-in-General with systematic completeness. Such a *mathesis universalis* may be possible or impossible, but the kind of mathematics we are discussing is not of this sort. The requirements in this formulation of a *mathesis universalis* are so enormous that we will only become conscious of the implications if we imagine the individual requirements concretely. Mathematics is supposed to be a determinate plurality in its various subdivisions as well as in its totality, hence a consistent field. But it is all too well-known that we do not know of a single subdivision of mathematics whose consistency can be proved or assumed. According to Husserl, however, not only is a single such field (axiom or system of theorems) supposed to exist, but there is even a system of such systems. The most interesting part of such a system would unquestionably be the system of all possible axiom systems—simply an unattainable goal, it seems to me. What is more, this system of all possible axiom systems is supposed to be the system of all possible variations of Something-in-General. If the concrete requirements are compared with those of *Principia Mathematica*, it is clear that in no sense are these requirements fulfilled in that work. To begin with, there is no proof in *Principia Mathematica* that the field dealt with there is a determinate plurality; its consistency is never once proved. Moreover, it cannot be read as a system of all possible axiom systems; a systematic account is not even provided for the small number of axioms used in this work. Thus it fails to give the system of the variations of Something-in-General in part any more than it does as a whole. *Principia Mathematica* is not a system at all, not to speak of its being able to carry out the excessive demands of a system of all systems.

Husserl discusses the concrete example of $7 + 5 = 12$ in the *Philosophie der Arithmetik* where he rejects the views of F. A. Lange and sides with those of Robert Zimmermann as presented in "Kants mathematischen Vorurtheil."[106] The definition given in the *Logische Untersuchungen* of analytic and synthetic propositions is a development of Bolzano's distinction and so ultimately has in common with Kant only the use of certain words.[107]

Hermann Ritzel (1880–1915), a student of Husserl's who fell in World War I, comes to a very unusual result in an otherwise fine and clear work. He comes to the conclusion that the judgment $7 + 5 = 12$ is analytic in scientific thought and synthetic in nonscientific. If there is no support to be found in Kant for making relative the distinction between analytic and synthetic judgments, such a sliding between *scientific-nonscientific* is entirely foreign to all Kantian thought. Such a distinction may be valuable for the views Ritzel presented; Kant's considerations remain untouched by it.

It must be admitted from the outset that the numbers can be defined in Couturat's way, for 5 is $4 + 1$ and 6 is $5 + 1$. But if the meaning of the numbers is so stipulated, then the equalities must be regarded as analytic judgments, more precisely, in fact, as tautological. . . . Taking suitable definitions as a basis, the proposition $7 + 5 = 12$ can therefore be regarded as an analytic judgment, and with good reason. The question arises, however, of whether the meaning of the number concepts defined in this way is also their usual meaning or whether we do not mean something different by the numerical expressions in a somewhat nontechnical use, so that the judgment is no longer analytic for this use of words. In fact, this is how matters stand.[108]

5. THE MARBURG SCHOOL

The school of Marburg does indeed call mathematical judgments synthetic, but we will see that this designation is not really made in the original Kantian sense. The main works of this school are *Kants Theorie der Erfahrung* (Kant's Theory of Experience) of 1871 and the *System der Philosophie* of 1889–1904 by Hermann Cohen (1842–1918), *Die logischen Grundlagen der exakten Wissenschaften* (The Logical Foundations of the Exact Sciences) of 1921 by Paul Natorp, and *Das Erkenntnisproblem in Philosophie und Wissenschaft der neueren Zeit* (The Problem of Knowledge in Philosophy and Science in Modern Times) of 1906–1920 and *Substanzbegriff und Funktionsbegriff* (The Concepts of Substance and Function) of 1910 by Ernst Cassirer (1874–1945).

It is well-known that Cohen started his interpretation of mathematics with the infinitesimal calculus in his *Das Prinzip der Infinitesimalmethode und seine Geschichte* (The Principle of the Infinitesimal Method and Its History) of 1883.[109] This approach inevitably forces the whole interpretation to become Leibnizian, and because of this, perception as independent source of knowledge is summarily rejected. Paul Natorp, for example, says:

> The following philosophy, stemming from Kant, and also the present direction, which is just as much *orthodox* Neo-Kantianism, has taken more and more exception to the dualism between pure perception and pure thought and finally decided to break with it.[110]

Accordingly, a purely logical foundation of mathematics is sought. Right at the start, this approach affects the concept of number itself.

> We can only hope to get such a thing as long as we make the primary assumption that the numbers simply are, and once the concept of them is established as just what is established, they remain established, uncreated, imperishable, unchangeable.[111]

In such a purely logical foundation of mathematics no unprovable or un-proven principles are permissible. In accordance with this, Natorp offers proofs, especially of geometrical principles; he is also convinced of the provability of the principles of arithmetic:

In dealing with addition, I have expressly omitted the derivation of the particular laws of arithmetic for these operations, the commutative, associative, and distributive laws, because the derivation does not present any special difficulties.[112]

We can see in Natorp's 1901 article "Zu den logischen Grundlagen der neueren Mathematik" (On the Logical Foundations of the New Mathematics) that he follows Grassmann's method of proof.[113] There is no doubt that this derivation of the whole of mathematics from a single origin is wholly in accord with Leibniz's way of thinking.

Such an attitude to mathematics leads, however, to a completely unsatisfactory interpretation of the Kantian synthesis. This can be seen in Cohen's *Kants Theorie der Erfahrung* because the whole problem of synthetic judgment is hardly dealt with at all. Cohen did realize this, but he makes a virtue out of by explaining at the beginning of his preface to the first edition: "In this work I have undertaken to give a new foundation to the Kantian theory of the a priori."[114]

But Kant had formulated the main problem of the *Critique* as the question, "How are synthetic judgments a priori possible?" Thus science, in particular, mathematics, is interpreted in two ways, as a priori and as synthetic. The fundamental value of Cohen's work for the understanding of Kant should never be underestimated, but in the long run, interpretation of the a priori exclusively proves to be inadequate. Accordingly, the discussion is scanty.[115] The interpretation Cohen himself gives of analytic and synthetic judgment fails to grasp Kant's intent.[116] Namely, Cohen maintains that synthetic judgment is of a kind which stays within the bounds of experience and is thus legitimate, while analytic judgment, on the contrary, goes beyond the bounds of experience.

Cohen evidently noticed the inadequacy of this interpretation when he wrote somewhat sadly:

> So the result of this inspection of the Kantian examples is that it does not contradict the meaning of the distinction which it is supposed to make clear; but, so far as examples can measure up to concepts, they are appropriate.[117]

It is certainly a sorry result when an interpretation can do no better than not directly contradict Kant's own examples, and it does not seem to me to be satisfactory to explain that the meaning is in agreement but that Kant's examples are poorly chosen. Moreover, the proposition $7 + 5 = 12$ is not even an example of an arithmetic proposition but instead exemplifies the fact that addition is a fundamental procedure of arithmetic. So it is not that Kant offers an example but that he means to demonstrate the correctness of his views by looking at a fundamental procedure.

Although Natorp, too, speaks of Kant's "not always fortunate choice of 'example,' "[118] he does try to give a more detailed interpretation of synthetic judgment:

> With this Kant distinguishes the stage (*Stadium*) of setting the problem sharply from the stage of solving it. . . . But that is exactly right. If I write $1 - 1$, $1-2$, $1:1$, $1:2$, and even $0:0$ or $\sqrt{2}$, $\sqrt{-1}$, or the formula of an infinite series, and so on, there are then relations which if otherwise already known, also fall under terms already stated, but stated at first only in the sense of setting problems, which are not thereby solved or even solvable. The most accurate understanding of what is required does not tell whether the problems are solvable at all or, if they are, whether in one or more or even in infinitely many ways. Only a further development of the arithmetic operations already known, which first has to be proved, can decide this; indeed, it can become necessary to introduce an entirely new kind of arithmetical operation, e.g., the method of limits and with it differentiation and integration. Zero and the negative numbers were not given in the original series of numbers beginning with 1, the fractions not with the integers, the irrational not with the rational, the imaginary not with the real. If the new kind of number were to develop from an impossible arithmetical concept into a possible and real one, an entirely new creation would be required, which perhaps for centuries no one has tried; nor would anyone have ventured to do so without the imperative need of the unlimited, universal performance of arithmetical operations already known.[119]

Natorp follows Hankel in this and sees the essence of mathematics, in particular the essence of its synthetic character, in the principle of the permanence of the formal laws. We have already seen that there is no concrete support for this idea to be found in Kant. It is, however, possible to develop it out of the Kantian approach, namely, from the approach to symbolic construction; but the principle can never suffice as a comprehensive interpretation of synthesis. By abandoning perception, the Marburg school has cut itself off from any real penetration into the problem of the Kantian synthesis.

Summary

We have established that perception, symbolic construction, and expansion are the essential characteristics of the Kantian synthesis in arithmetic. Perception finds its fundamental expression in axiomatics. I hope I have clarified the significance of the statement of the axioms of arithmetic by Schultz and what part is to be credited to Kant. Time has a fundamental importance in all of these questions. The pure perception of time is the source of the axioms; the symbolic construction of arithmetic is carried out in time, and finally, the capacity for expansion has a temporal character. All the real attacks directed against synthetic judgment in arithmetic are against time as the foundation of the Kantian theory of mathematics. Bolzano, Brentano, Husserl, Cohen, and Natorp all assume that mathematical propositions, in particular those of arithmetic, exist in themselves.

It will also be necessary to go through Kant's account of geometry in all its details. Then we will finally have all the material for making clear the fundamental importance which Kant attributed to time for mathematics.

Appendix
Notes
Bibliography
Indexes

Examination of Kant's "Critique of Pure Reason,"
Part I, Section 4
by Johann Schultz

Translator's note:

Johann Schultz (1739–1805) was both a pastor and a professor of mathematics at the University of Königsberg. He became a privatdozent at the University of Königsberg on the basis of his habilitation dissertation, "De geometria acustica seu solius auditus ope exercenda," in 1775, and gave lectures in pure mathematics and astronomy. He received a chair as professor of mathematics in 1787, delivering as his Inaugural Dissertation, "De geometria acustica nec non de ratione 0:0 seu basi calculi differentialis." Having been a pastor at several churches as well, he became the Hofprediger (court chaplain) at the Schlosskirche (palace church) in 1786. He was a friend and follower of Kant's and wrote works explaining and defending the Kantian philosophy. (See the bibliography.) The *Prüfung der Kantischen Critik der reinen Vernunft* (Examination of Kant's Critique of Pure Reason) was supposed to cover the entire *Critique of Pure Reason*, but Schultz finished only the first two parts, the first in 1789 about the Introduction and the second in 1792 about the Transcendental Aesthetic. He deals with the distinctions between the analytic and synthetic, a priori and a posteriori judgments, and Kant's theory of time and space; he defends Kant's theory in detail against the criticisms of such men as Johann Feder, Gottlob August Tittel (1739–1816), author of *Kantische Denkformen oder Kategorien* (Frankfurt am Mayn, 1787), J. C. F. Vornträger, author of *Ueber das Daseyn Gottes in Beziehung auf Kantische und Mendelssohnische Philosophie* (Hannover, 1788), Adam Weishaupt (1748–1830), author of *Zweifel über die Kantische Begriffe von Zeit und Raum* (Nürnberg, 1788), Johann Albert Heinrich Reimarus, author of *Ueber die gründe der menschlichen erkenntniss und naturlichen religion* . . . (Hamburg, 1787), and J. A. Eberhard. To quote Otto Liebmann's article in the *Allgemeine Deutsche Biographie*, "His indubitable superiority in this dispute depends above all upon his thorough knowledge of elementary and higher mathematics."[1]

The following section about arithmetic is one to which Gottfried Martin makes frequent reference. The page numbers of the original appear in brackets.

Part I, Examination of the Introduction
Section 4, Are there theoretical sciences which contain synthetic judgments a priori?[2]

III. All Pure Arithmetic and General Mathematics Is Actually, Like Geometry, a Totally Pure Science Consisting of Pure Synthetic Propositions A Priori.

The *magnitude* or *quantity* of a *thing* is its inner determination which is generated from the combining of the homogeneous. The thing itself that has quantity is called a quantity *in concreto* or a *quantum*. Hence a thing is a *quantum* insofar as there is a combining of the homogeneous in it. The science of *quanta* as such is called *arithmetic, mathesis*, or *mathematics*. That part which is concerned with quantity in general without reference to any particular kind is called *general* mathematics (*universalis*); the whole of the remaining part which has special given kinds of quanta as its object is called *special* (*specialis*). Geometry and trigonometry therefore belong to *special* mathematics, for space, with which [212] they are concerned, is already a special kind of quantum. So the practical art of calculation and, in general, the whole of applied mathematics belongs to it. Since special mathematics already deals with determined qualities, determining their quantity necessarily has reference to their quality. The geometer, accordingly, examines the *similarity* of figures when he examines their quantity, i.e., the identity of their qualities, and he would, for example, try in vain to determine the magnitude of a line if he did not want to see whether it was straight or curved, and what kind of curvature it had.

In contrast, *general* mathematics abstracts completely from the different qualities of quanta, so it deals only with quanta as such and their *quantity*, and it only examines all *the possible ways of combining the homogeneous*, by which the magnitude of a quantum in general is generated and can be determined. Accordingly, the determination of the magnitude of every quantum, of whatever kind, depends on it, so it is the basis of every special kind of mathematics.

The two main ways of generating quantity by combining the homogeneous are *addition*, where one asks for the magnitude which is generated when one combines a given quantum with another homogeneous one; i.e., how large is x, when $x = a + b$? [213] and *subtraction*, where one asks for the magnitude by means of which another given is generated through combining with a given homogeneous; i.e., how large is x, when $x + b = a$? In this latter case, one calls the *relation* of the given magnitudes a and b *arithmetical*, and the x sought for, its *Name* or *Difference*. Thus addition and subtraction, together with what is based on these arithmetical relations, are the two most universal ways of generating magnitude which hold for all quanta, and so constitute the *Mathesin universalissimam*. And since subtraction is nothing more than the addition of the sought-for quantity x to the given b, addition is the real basis of all mathematics. Since, however, in generating the quantity x through addition or subtraction, the given quanta *a* and *b*, insofar as they are homogeneous (i.e., regarded as quanta), can be anything; but apart from this, it is impossible to determine how the quantity of the one is generated from the other; so something does indeed always remain undetermined in generating the quantity *x* through addition and subtraction, and one gets only a new *quantum* through it from the given quantum; yet the concept of its *quantity* still remains totally undetermined. Suppose, for example, I pour one heap of wheat onto another. Then the addition of these heaps, *A* and *B*, gives me a new one, *C*; [214] but since I know only the given ones as quanta that are independent of each other without knowing how the quantity of the one was generated from the other, I also do not know what kind of quantity has resulted from their combining; the addition of the given quanta merely gives me a new quantum, but not a concept of their quantity. So if we should want a determinate concept of the quantity of a quantum *x* which is generated through the combining of something homogeneous, it is first necessary that the *same* homogenous quantum *a* be continually added to itself and so $x = a + a + a + a + \ldots$. Then *x* is called a *multiple* of *a*, and *a* an *aliquot part* of *x*, and also the *unit of measure*. And this leads to two new ways of generating magnitude, namely, *multiplication*, where one seeks the multiple from a given quantum, and *division*, where, the other way round, one requires an *aliquot part* of it. Second, because the concept of the multiple and of its aliquot parts is but a totally *undetermined* concept of *plurality* (*Mehrheit*) or *manyness* (*Vielheit*) which does not show how many times the unit [*Maass*] *a* is contained in the multiple *x*, this *how-many-times* must still be *determined*. As soon as this is done, we know how many times the unit *a* or an aliquot part of it has to be taken, if the quantum *x* is to to be generated [215] then we have a *determinate concept* of the magnitude of the latter. We then say, "We have measured the quantum *x* by the quantum *a*, or have determined the *geometrical* relation between *x* and *a*." The determination of the *how-many-times*, which consists of combining the *many into a whole*, is called a *number*. So a determinate concept of the *magnitude* of a quantum, i.e., the measurement of it, is only possible through *numbers*. Now the science of numbers is called *arithmetic*, and this includes both the *elementary* and the *higher* parts, i.e., *algebra* together with the *differential* and the *integral calculus*. So general mathematics, except for the short theory of general addition and subtraction which forms its first foundations, is just the elementary and higher arithmetic or science of numbers.

That this is, first of all, not an empirical but a pure science a priori, cannot be doubted, since all its propositions have absolute necessity and so also the strictest universality and apodictic certainty. The arithmetician is not only convinced of the universality of his propositions; he is also perfectly conscious that his conviction is based not on perception and induction but on proofs independent of all perception. No science [216] offers more opportunities to be acquainted so vividly with the amazing difference between demonstrations a priori and proofs which depend merely on induction than arithmetic—and especially algebra and the differential calculus. The certainty of the universality of a proposition can, in fact, be established to a very high degree merely by induction, because one can check the correctness very easily from all possible

sides by innumerable examples. A multitude of important algebraic propositions, e.g., the binomial theorem, various sums of series, Harriot's theorem that
every equation has as many positive roots as changes of sign and as many
negative roots as there are sequences of the same sign, etc., and so almost the
whole analysis of the infinite—all this was actually first discovered by induction.[3]

Yet it was quite well known that so long as one was not in the position to
demonstrate these propositions a priori, certainty about them was not apodictic
but merely empirical, and this certainty was based on examples because the
propositions had always been found correct in application. Apart from this,
however, it is immediately obvious from the nature of general mathematics that
it cannot have an empirical origin. For since geometry has a special given [217]
quantum as its object, namely, space, it is indeed necessary to inquire whether
its object is something empirical. Only general mathematics has *no* special
quantum at all as its object, and instead abstracts completely from any quantum;
it deals only with the *quantity* of a thing in general, or with *number*, and so never
has anything sensible as its object, let alone anything empirical, but only the
purely intellectual concept of quantity, i.e., unity, plurality, and universality.
Thus it would be an obvious contradiction to hold even one of its propositions to
be empirical; it therefore consists of pure propositions a priori.

In fact, I doubt that any empiricist could really make such a mistake as to hold
that a properly demonstrated arithmetical or algebraic proposition is empirical,
and, for example, persuade himself that "two times two is four" is only certain
because he has not yet perceived anything different—that is, so long as he does
not actually maintain that the law of contradiction itself is empirical and is
certain only insofar as one can expect that, having been proved by all past
experience, it will continue to hold in the future. Rather, the opponents of our
philosopher [Kant] unanimously say that all propositions of arithmetic are
analytic, and thus [218] that they are propositions a priori, since it would be
ridiculous to appeal to experience in merely analyzing concepts. In fact, I could
not be less surprised by any other objection than this one, for the idea that the
propositions of arithmetic, e.g., $7 + 5 = 12$, are merely analytic, are indeed
completely identical, is so plausible that here I really admire the acute insight of
our Kant for having penetrated so deeply into this apparent truth. So far as I can
tell, the investigation of the real nature of arithmetic is precisely one of the most
difficult problems. But then it is all the more important to bring out the truly
synthetic nature of this science as clearly as possible. The main reasons for this
are as follows:

1. Arithmetic Has Both *Axioms* and *Postulates*, Exactly Like Geometry.

It does, indeed, at first appear as if there were no axioms of arithmetic. For all
the propositions usually presented as axioms in textbooks of arithmetic are
either purely analytic consequences of definitions, e.g., like the propositions
that the whole is equal to its parts taken together and that no part is greater than
the whole; or else they are consequences of the proposition $A = A$, i.e., of the
principle of contradiction itself, e.g., like the proposition that two quantities
which are equal to a third are equal to each other; or they [219] can actually be
proved, e.g., like the propositions that equals added to equals, or subtracted

from them, yield equals, as *Wolf* [sic] has already shown in his *Elementa matheseos universae*.[4]

Nonetheless, there are, in fact, two axioms at the basis of all of elementary and higher arithmetic, namely, the following:

FIRST AXIOM OF ARITHMETIC

The *quantity of the sum is the same* whether one adds the *second* to the *first*, or the *first* to the *second*, i.e., it is always the case that $a + b = b + a$.

SECOND AXIOM OF ARITHMETIC

The *quantity of the sum is the same* whether one adds to a given quantum another one at once as a *whole* or by *each of its parts, one by one*, i.e., it is always the case that $c + (a + b) = (c + a) + b$.

These propositions are absolutely necessary for arithmetic. For to add a number 5 to another one, 7, means to add onto a number 7 the whole 5, i.e., all its units taken together at one time and thereby to generate a single number as sum. But if I merely stick to this concept of adding, the desired sum can never be generated, for I can never make the least out of the concept of $7 + 5$ or [220] $7 + (4 + 1)$ or $7 + (1 + 1 + 1 + 1 + 1)$ in itself. So I simply go beyond this concept and instead of adding the whole 5, or its units taken together at one time, onto the number 7, I have to combine the 7 successively with it by each unit, and instead of setting $7 + (4 + 1)$, first set $7 + (1 + 4)$, and instead of this, $(7 + 1) + 4$. So then because of the concept of 8, since it is $7 + 1$, I get $8 + 4$, i.e., $8 + (3 + 1)$. But then instead of this I must again set $8 + (1 + 3)$, and for this $(8 + 1) + 3$, so I get $9 + 3$, i.e., $9 + (2 + 1)$. For this, $9 + (1 + 2)$, and setting $(9 + 1) + 2$ for it gives $10 + 2$, i.e., $10 + (1 + 1)$, and setting $(10 + 1) + 1$ for this finally gives $11 + 1$, i.e., 12. Every arithmetician knows that this is the only way we can gain insight into the truth of the proposition that $7 + 5 = 12$. But how do I know that this arbitrary procedure, which is not part of the concept of addition at all, does not make any difference to the magnitude, does not alter the quantity of the sum of $7 + 5$, namely, first, that $4 + 1 = 1 + 4$ always, or in general, that $a + b = b + a$, and secondly, that $7 + (1 + 4) = (7 + 1) + 4$, or in general, $c + (a + b) = (c + a) + b$? Clearly, no further proof of this is possible, because a proof could only be derived from the concept of the way one gets the sum of many [221] numbers, so every proof sought for must obviously be a *petitio principii*. Every careful mathematician sees this only too well in discussions of these two theoretical propositions. Since they are nevertheless apodictically certain, they are true axioms, for without them the certainty of addition and so of the whole of arithmetic and even of the proposition that two times two is four, would disappear.

But besides these two axioms, arithmetic, like geometry, also has genuine postulates or practical basic propositions upon which its whole possibility depends, no matter how much they have been overlooked up till now. They follow:

FIRST POSTULATE OF ARITHMETIC

To generate the concept of *one* quantum from *many* given homogeneous quanta *through their successive combining*, i.e., to transform them into a *whole*.

SECOND POSTULATE OF ARITHMETIC

To increase, and to decrease, any given quantum *as much as one wants*, i.e., *to infinity*.

The first postulate deals with what the entire *possibility* of a *whole* or a *sum* depends on, and so with the very possibility of all *addition*, for it really [222] says no more than *to add several quanta*. Admittedly, this proposition is usually posed as a problem; however, the rule given for its solution obviously presupposes the way it has to be done as a self-evident postulate. The solution given for addition in general is: in order to determine the sum x of the quantities a and b, set $a + b = x$. But in itself this solution is no more than a mere characterization of it which simply expresses the problem which was presented in words in the more convenient language of symbols without producing the least concept of the sum x sought for, or even indicating a possible way of finding it; but it is clear that the solution does presuppose the concept as being possible when a and b are given as *objects of perception* [*Anschauung*], because the way *these can be unified in one whole* is immediately evident from the perception of them itself. This is even clearer in the addition of numbers, for the solution prescribed for adding the one, the ten, the hundred, etc., each one separately, shows that it is merely designed *to make the addition of the larger numbers quicker and easier by using* the decadic system. This is because it shows how conveniently one can also add many and large numbers [223] as soon as one merely understands how to add a given number to another one not over 9. If I should manage this and, for example, add 5 to 7, I have to imagine the *units* out of which the number 5 is composed, according to the series *individually*; then I have to *add one after the other* onto the number 7 and so generate the number 12 as a whole by means of successive combining. But there can be neither a special rule nor a proof that or how this process is *possible* in general, yet everyone has an immediate and certain idea of it, and thus the practical judgment which reads: to every given number to add every other, apart from the short-cut rules with large numbers, is not a problem but a *postulate* just like the postulate: to draw a straight line from one point to another.

Since general mathematics is not concerned with a particular quantum but only with the *quantity of a quantum* in general, the second postulate says that, be a given quantity ever so large or small, a greater one, or a smaller one, can always be thought of. This proposition is just as undeniably certain taken in the greatest generality, not only with regard to finite but also to every infinite quantity. Not only every finite quantity can be decreased or increased without end, but the algebraist can even regard an endless continuous series of units as a [224] whole $= \infty$, and can take this again n times or ∞ times, and so generate in thought $n \times \infty, \infty^2, \infty^3, \infty^n, \infty^\infty$ etc. [5] But this proposition is at the same time *immediately evident* so that no proof of it is at all necessary or possible. Therefore, it is likewise a *postulate*.

When one compares these *arithmetical* axioms and postulates with *geometrical* ones, one is struck by their remarkable similarity. Two arithmetical *axioms* taken together say: *No more than one sum is possible from many given quanta*; and the straight line axiom corresponds exactly to it: No more than one straight line is possible from one point to another. And the two geometrical postulates obviously correspond to the two *postulates* of arithmetic: To draw a straight line from one point to another, and: To extend every straight line without end.

[225] Arithmetic thus has true axioms and postulates. Now axioms and postulates are always synthetic propositions, as shown above in II, 2,[6] and so addition is always synthetic. Now all elementary as well as higher arithmetic depends on it because all possible ways of calculating are ultimately addition, the concept of *number* itself, indeed, being generated merely by the addition of units. So clearly all of arithmetic or general mathematics consists only of *synthetic* propositions a priori throughout.

Moreover, since no proof of postulates and axioms is possible, the combination of the predicate with the subject in them does not depend on a concept, but on *direct acquaintance* [*unmittelbare Vorstellung*], i.e., on *perception [Anschauung]*. Therefore all of arithmetic or general mathematics depends on *perception*, just like geometry; in particular, since it is a pure science a priori, it depends on *perception a priori*.

The truth of this can be seen quite clearly from the nature of its axioms and postulates and from the concept of counting itself. For example, when I add 5 to 7, I first have to think of the units of the number 5 singly, thus: $1 + 1 + 1 + 1 + 1$. Now every one of these units, the *quantity* as well as the *quality*, is in itself completely homogeneous, and so the intellect cannot distinguish [226] them by the slightest *internal* mark, and so not by a concept. So this is possible only by *external* or *sensible* marks, i.e., by *perception* and not concepts. Thus if the intellect were able to think of the *units* of which a number is composed as *several different* things, it would simply have to call on some kind of *perception* for help, for without this the concept of their *plurality* would be totally impossible, since it would have to think of all of them together as *eadem numero*, i.e., as *just one*. Now we cannot distinguish things perfectly homogeneous with respect to their internal marks, and so not distinguishable at all by intellect alone, except by our imagining them in *different locations of space* or at *different points of time*. So the *thought* of the several units which a number contains is only possible by our *thinking* of them in different locations of *space*, or of *time*, or of *both together*. But it is easy to see that only the last occurs here. For, first of all, it is impossible to comprehend the several units which a number contains except as *successive*, i.e., as a series in time. But since time flows and continually changes the imagination [*Einbildungskraft*] would not be able to grasp these units at all and hold them fast without attaching them to something *firm*, [227] portraying them so it could always run through them again at will and present them by direct acquaintance. So in order to think of the several units which a number contains, it is necessary to think of them in space as well as in time. This is most clearly confirmed by what we really do when we want to think distinctly of the individual units contained in a number. For we set them all *successively* as separated *points*, or as *lines in space* next to each other, and it is simply impossible for the algebraist to think of several infinite series of numbers without thinking of them proceeding *one after the other* in different *directions in space*. Nevertheless, to determine the sum of 7 and 5 it is not enough for me to set the units of the number 5 individually in space, but I must also join them all to the number 7 in such a way that a whole, i.e., the number 12, is generated. This combining [*Verknüpfung*], however, is possible only if I add each *individually, one after the other,* to the number 7, and consequently undertake to combine them in a whole successively, i.e., in time, namely, $7 + 1 = 8, 8 + 1 = 9, 9 + 1 = 10, 10 + 1 = 11, 11 + 1 = 12$. Generating the concept of number itself requires something similar, for here I can only experience the combining of units successively [228] so that I first generate a whole which I call 2 from 1 + 1, a whole which I call 3 from 2 + 1, a whole which I call 4 from 3 + 1, etc. Determining *quantity*, then, requires that the homogeneous from which it is to be generated can be presented in a spatial image. But what is essential, that on which the very possibility of its generation directly depends, is that we can think of several homogeneous things *successively*, hence the idea [*Vorstellung*] of time. Thus, the immediate certainty of both *postulates* of arithmetic depends on it alone: The mere consciousness [*Bewusstseyn*] that we can combine several imaginable units in space *successively* to each other, i.e., in time, shows us directly *that* and *how* the generation of a *totality* [*Allheit*] or a *whole* [*Ganzen*] is always possible from the concepts of *unity* and *plurality*, and since time is infinite and eternal, that we may never allow the generation of totality to *stop*, neither forward nor backward. The certainty of the arithmetical *axioms* depends just on this. For that I can set $a + b = b + a$, i.e., that the *value* of $a + b$ is the same whether I think of a or b *first*, likewise that the quantum $a + b$ added to c gives the same value whether I add the whole $a + b$ together and at one time to c or successively, only one part after another—this depends only on the consciousness that the *quantity* is an *internal property* of a quantum, that is, [229] a *concept*, while time simply belongs to its external properties and is solely *perceived*; hence it behaves indifferently to quantity as a concept, since merely perceiving a thing cannot alter the *concept* of it in the slightest. Only the intellect has formed the concept.

2. The whole of arithmetic is based on *equations* [*Aequationen*] or *equalities* [*Gleichungen*]. Now at first glance, an equation, e.g., $7 + 5 = 12$, has very much the appearance not only of an analytic proposition but also of an identity in which the subject and the predicate are one and the same concept, or even that one is the true name of the other. Hence people usually explain the equality $7 + 5 = 12$ thus: $7 + 5$ *is* 12, or even, as H. Reimarus[7] would have it: $7 + 5$ is *called* 12. But if this were correct, then it would not be necessary to add at all, for anyone who simply knew the number would have to think of the concept of the number 12 immediately in the concept of $7 + 5$, so that it would be as impossible not to think of the number 12 when thinking of $7 + 5$ as it is for us to think of the completely perfect being without thinking of God. And since the whole of general mathematics is basically only addition, he would also have to recognize in the [230] combined [*zusammengesetzten*] equalities, e.g., $4\sqrt[3]{3} + 8^4 - 7^3 = x$, the value of the predicate x follows immediately from the given subject without any calculation. Furthermore, since the number 12 can also be the predicate in innumerably many other equalities, e.g., $18 - 6 = 12, 3 \times 4 = 12, \sqrt{144} = 12$, etc., it is also the case that $7 + 5 = 18 - 6 = 3 \times 4 = \sqrt{144}$, etc. So $7 + 5$ and $18 - 6$, 3×4, $\sqrt{144}$, and innumerable other

combinations of the same sort would *only* be *one* concept, and I would then have to think of all these as well whenever I think of 7 + 5.

The underlying mistake here is confusing the sheer sameness [*Einerleyheit*] of *quantity* with the sameness of the *concept of quanta in general*. For an equation expresses merely the *equality* of quanta, i.e., the *sameness* of their *quantity*, however *different* the *concepts* of the quanta themselves may otherwise be. So, for example, two *triangles A + B* together can be equal to a *circle*. The true sense of an equation, e.g., 7 + 5 = 12, is thus just this: the *quantity* of 7 + 5 is the *quantity* of 12, or the *quantity* of 7 + 5 and the *quantity* of 12 are the same, although I do not think of the *concept* of 7 + 5 at all when I think of the *concept* of 12. But now the question arises: How do I know that in the proposition: the [231] quantity of 7 + 5 is the quantity of 12, that the predicate belongs to the subject? If the proposition were analytic, I would have to be able to recognize this from the mere analysis of the concept of 7 + 5. By itself, this concept says no more than that the *whole* 5 is added on to the number 7 *at one time*. Granted, therefore, that I also wanted to analyze the number 5 into the parts 4 + 1, or 1 + 1 + 1 + 1 + 1 which *taken together* are equal to it; then I would now have the concept 7 + (4 + 1), or 7 + (1 + 1 + 1 + 1 + 1) instead of 7 + 5. But I still know the *quantity* of this just as little as I knew the quantity of 7 + 5, and the mere analysis is completely finished here, as already shown. Thus, if the proposition 7 + 5 = 12 were *analytic*, it would be totally unprovable [*unerweislich*]. Consequently, it would have to be immediately evident in itself, without even any future analysis, and so it would have to be a real axiom; and for precisely that same reason, all propositions of the whole of arithmetic and algebra would also have to be pure axioms, whose truth everyone would immediately have to recognize. Thus it is evident that the mere analysis of the concept of 7 + 5 can never lead me to the conclusion that its quantity is as much as that of 12; but if I am to recognize this, I simply have to go beyond this concept and first use the two axioms of arithmetic for help, together with the first postulate. The axioms teach me that I am free to add [232] the parts of 5 to 7 successively instead of adding it all at once as a whole, and thereby at the same time changing the order of the parts. The postulate tells me that and how this is possible. So I first come to the insight that the quantity of 7 + 5 is the same as that of 12 through this laborious synthetic procedure. Thus the proposition 7 + 5 = 12, and every equation B = A generally, is manifestly *synthetic* with the sole exception of A = A.

3. Analytic judgments *cannot enlarge* [*erweitern*] our concepts at all but can only *elucidate* [*erläutern*] them and *make them clear* by the analysis of their components, for they only pick out a concept in the predicate which is already in the subject. Now *general mathematics* is a science which is capable of such astonishing expansion merely a priori that no other rational science can be named its equal in this respect. Hence the development of the whole of *special* mathematics depends for the most part only on the expansion of the *general*. Thus it is an obvious contradiction for a science of such enormous scope to arise merely by analyzing the concepts of unity, plurality, and totality; so this fact alone shows that it must be synthetic throughout.

[233] In this way it is certain that arithmetic or general mathematics is a

totally pure synthetic science a priori just as geometry is, and that this is therefore true of the *whole of pure mathematics* [*Mathematik*]. Since, further, all applied mathematics is based on the pure, it is also self-evident that it also depends from beginning to end on synthetic propositions a priori, and that whatever has apodictic certainty has to owe it solely to these propositions.

To construct a concept means to represent it in pure perception [*Anschauung*], i.e., perception a priori. Now since all synthetic propositions, not only those of geometry but also those of arithmetic, must, as has been shown, be represented in perception a priori insofar as we want to understand why they are possible and correct; the Kantian contention that mathematical knowledge consists not merely of deductions from concepts but also of the *construction of its concepts* is established beyond all doubt.

Furthermore, we cannot perceive by our intellect but can only think with it, and all that we are directly acquainted with belongs not to the intellect but solely to our sensibility, yet all objects which mathematics is supposed to be applied to must be capable of [234] being presented. It then follows that mathematics is applicable only to *sensible* objects, and thus that hardly a greater absurdity can be imagined than subsuming merely *intellectual things* [*Verstandeswesen*] or *things-in-themselves* under calculus or geometry, and, for example, dreaming of a mathematics of simple and spiritual substance, or indeed of a *Geometria divina*.

Utimately the axioms and postulates of arithmetic are founded not on concepts but on perception a priori, and the one they are founded on is none other than the *successive* in the connecting of units, and thus is the idea [*Vorstellung*] of *time*. It is thus clear that what we are directly acquainted with a priori (and upon which arithmetic is essentially based) is none other than *time*. So it is obvious that time is no more a *concept* of the intellect than space is, but is a *sensible* idea or *perception*. It is not empirical but *pure, a priori* and so it is the second pure and necessary form of our sensibility and indeed the form of inner sense without which no *perception of the sequence of our thoughts* would be possible. So we can conclude with certainty that all objections which have been made of this Kantian position, and may yet be made, [235] depend upon a misunderstanding. I must present the following in order to show that they really depend upon it. But first I merely want to note that I have been fortunate enough to discover the real axioms and postulates of time. Here they are:

AXIOMS OF TIME

1. Between two given points of time there is *only one* time.

2. All parts of time are *similar* to each other, and so two *equal* parts of time are *congruent*, i.e., equal and similar.

POSTULATES OF TIME

1. Between two given points of time *there is always* a time.

2. Every given part of time can be *extended on both sides*, i.e., *forward and backward, without end*.

These axioms and postulates of time show the most exact correspondence to the axioms and postulates of the straight line, and it is clear why time can only be represented spatially as a straight line. But [236], indeed, the similar origin and nature of the ideas of *space* and *time* also reveal themselves from this. Our philosopher has, indeed, also had the question posed to him, among others: Why is it that perceptual nature of time has hardly helped us at all to get one or more of the propositions, while that of space has, even to get a whole science, that of geometry?

This, however, overlooks that in point of fact we do have an *entire pure science of time* which is of no small scope, namely, *pure mechanics*, whose main objective is, above all, the measurement of time. It is more a cause of wonder that the science which has time as its object really has so great a scope, since time has but a single dimension and also merely corresponds to a *straight* line, and, what is more, only to a *single*, infinite straight line.

Notes

Abbreviations Used in Notes

Ak. Immanuel Kant, *Gesammelte Schriften*. Edited by the Koniglich Preussische Akademie der Wissenschaften and by the Deutsche Akademie der Wissenschaften in Berlin. 23 vols., Berlin and Leipzig, 1900–55.

CPR Immanuel Kant, *Critique of Pure Reason*

CJ Immanuel Kant, *Critique of Judgment*

Ak. Gottfried Wilhelm von Leibniz, *Sämtliche Schriften und Briefe*. Edited by the Preussiche Akademie der Wissenschaften and by the Deutsche Akademie der Wissenschaften in Berlin. Berlin, 1923.

Translator's Preface

1. [tr.] Obituaries about Martin with additional information appeared in the two journals with which he was so closely associated: *Kant-Studien* (1972) and *Studia Leibnitiana* IV (1972).

Preface

1. [tr.] See the article about Johann Schultz by Otto Liebmann (1840–1912) in the *Allgemeine Deutsche Biographie*.

2. [tr.] Presumably Martin emphasizes this period since it is the period in which Johann Schultz published three works explaining and defending Kant's critical philosophy. In the present work, the relation between Schultz and Kant is central for Martin. We can assume that Kant also continued to be interested in mathematics between 1762 and 1780.

3. [tr.] German *Reflexionen*. These notes of Kant's are entitled *Reflexionen* in volumes XIV–XIX of the Academy edition of Kant's works. I have retained the spelling "Reflexion".

4. [tr.] Erich Adickes refers to the evidence for this in Richard Reickes' "Losen Blätter aus Kants Nachlass," of 1889, which I have not been able to consult. See Kant, Ak. XIV, Vorwort, vi. In 1936 Martin published "Herder als Schüler Kants. Aufsätze und Kolleghefte aus Herders Studienzeit" in *Kant-*

Studien, 41 (1936), 294–306. In 1964 Hans Dietrich Irmscher published some of Herder's notes of Kant's lectures during 1762–64. *Immanuel Kant. Aus den Vorlesungen der Jahre 1762 bis auf Grund der Nachschriften Johann Gottfried Herders.* Included are notes from two different lectures on mathematics (which Irmscher says might also be of the lectures of Friedrich Johann Buck, whose lectures on mathematics Herder attended as well as Kant's). They appear in vol. 6 of *Vorlesungen* in the Academy edition of Kant's works (Ak. XXIX), 49–66. On the basis of these lecture notes, Martin revised some of the views of the 1938 *Arithmetik und Kombinatorik bei Kant.* "Die mathematische Vorlesungen Kants," *Kant-Studien* 58 (1967), 58–62. However, the present work does not contain those revisions. I have included them in a note about the courses he gave in mathematics.

Introduction

1. Carl Friedrich Gauss, Ak. VIII, 170–74, in his critical book review of Johann Christoph Schwab's *Commentatio in primum elementorum Euclidis librum, qua veritatem geometriae principiis ontologicis niti evincitur, omnesque propositiones, axiomatum geometricorum loco habitae, demonstrantur* of 1814. Originally in the *Göttingische gelehrte Anzeigen*, 20 April 1816.

2. Wilhelm Franz Meyer (1856–1934), "Kant und das Wesen des Neuen in der Mathematik," *Zur Erinnerung an Immanuel Kant*, 309.

3. Hans Beck, *Einführung in die Axiomatik der Algebra*, Foreword.

4. [tr.] Martin uses *logisch* (*logical*) here, but *logicist* is the usual English term for the view he is referring to. E.g., cf. Rudolf Carnap, "The Logicist Foundations of Mathematics," in *Philosophy of Mathematics*.

5. [tr.] Martin should really mention Alfred North Whitehead (1861–1947) as well here since he was the coauthor with Russell of *Principia Mathematica*.

6. [tr.] Antoon Vloemans is mistakenly listed as "Vloomans" by Martin and as "Anton Vloomans" in some catalogues.

7. C. T. Michaelis, "Über Kants Zahlbegriff," *Wissenschaftliche Beilage zum Programm der Charlottenschule* (1884), 3.

8. Paul Mansion, "Gauss contra Kant sur la géométrie non euclidienne," in *Bericht über den 3. internationalen Kongress für Philosophie*, 439.

9. Louis Couturat, "La philosophie des mathématiques de Kant," *Revue de Métaphysique et de Morale*, 12 (1904), 340.

10. [tr.] *Losen Blätter* means *loose pages*. Since these notes of Kant's are entitled *Losen Blätter* in volumes XX, XXI, and XXIII of the Academy edition of Kant's works, I have left the original German untranslated.

11. Erich Adickes, *Kant als Naturforscher*, vol. I, 19.

12. Antoon Vloemans, *Anschauung und Verstand in der Entwicklung von Kants Theorie der Geometrie unter Berücksichtigung von Descartes, Leibniz und Gauss*, 120, 126.

13. [tr.] Note that Martin wrote this in 1938.

14. Friedrich Schmeisser, *Anleitung zum Selbstfinden der reinen Mathesis*, 105.

15. Emil Arnoldt, *Kritische Exkurse im Gebiete der Kant-Forschung* in *Gesammelte Schriften*, ed. O. Schöndörfer, vol. V, 177ff.

16. Friedrich Wilhelm Schubert, *Immanuel Kants Biographie*, in *Kants Werke*, eds. K. Rosenkranz and F. W. Schubert, Leipzig, 1842, vol. XI, pt. 2, 35.

17. [tr.] Martin revised his views about what he wrote here in 1938 in his 1967 "Die Mathematischen Vorlesungen Kants," based on the Herder lecture notes published by Irmscher. The gist of the article is as follows: Kant gave lectures on mathematics for sixteen semesters. The textbooks most often used were those of Christian Wolff. The German *Anfangsgründe aller mathematischen Wissenschaften* appeared first in 1710 and then in many later editions. (Kant owned the 1750 edition.) Shortly afterwards Wolff published a Latin edition, which is essentially the same as the German *Anfangsgründe: Elementa Matheseos*, 1713. This also went through many editions. Finally, in 1713 Wolff also published *Auszug aus den Anfangsgründen aller mathematischen Wissenschaften*, which shortened the *Anfangsgründe* by about a third. This too went through many editions. Kant owned the edition of 1749. He used the *Auszug* as the textbook for his course in the summer semester of 1758. Herder's notes show that Kant also used it in the winter semester of 1762/63, following it faithfully paragraph by paragraph. It is unlikely that Kant changed textbooks later. Unfortunately, Kant's own copy, which was listed in the catalogue of the auction of Gensichen's books (published by Arthur Warda), has not been found. Presumably, it would have Kant's comments on the text.

The *Anfangsgründe* begins with a systematic account under the title of *Kurzer Unterricht von der mathematischen Lehrart* and organizes the material into nineteen parts. The parts of the *Auszug* can be put together with Kant's courses in the following table:

Wolff's Auszug	Course SS 1761	Course WS 1761/62	Herder's Notes
1. Arithmetic	Arithmeticam	Arithmetices	Arithmetic
2. Geometry	Geometriam	Geometriae	
3. Trigonometry	Trigonometriam	Trigonometriae	
4. Mechanics	Mechanicam		
5. Hydrostatics	Hydrostaticam		
6. Aerometrics	Aerometriam		
7. Hydraulics	Hydraulicam		
8. Optics			
9. Catoptrics			
10. Dioptrics			
11. Perspective			
12. Astronomy			
13. Geography			
14. Chronology			
15. Gnomonics			
16. Artillery			
17. Fortification			
18. Architecture			
19. Algebra			

The course descriptions for the summer semester of 1761 and the winter semester of 1761/62 show that Kant gave a two-semester course then and that he covered the first three parts of the *Auszug* in the first semester and the fourth, fifth, sixth, and seventh in the second. The course descriptions for the summer semesters of 1756 and 1757, the winter semester of 1759/60, and the summer semester of 1761 show that the course must have been at least a two-semester one.

If the course covering arithmetic, geometry, and trigonometry is called *Mathematics I*, and the course covering mechanics, hydrostatics, hydraulics, and aerometry *Mathematics II*, and if we fill in the semesters about which we have no information or no information about what Kant covered with the corresponding courses which were running, we get the following series of courses.

1. WS 1755/56: Mathematics I
2. SS 1756: Mathematics II
3. WS 1756/57: Mathematics I
4. SS 1757: Mathematics II; Mathematics I
5. WS 1757/58: Mathematics II
6. SS 1758: Mathematics I
7. WS 1758/59: Mathematics II
8. SS 1759: Mathematics I
9. WS 1759/60: Mathematics II; Mathematics I
10. SS 1760: Mathematics II
11. WS 1760/61: Mathematics I
12. SS 1761: Mathematics II; Mathematics I(?)
13. WS 1761/62: Mathematics I
14. SS 1762: Mathematics II
15. WS 1762/63: Mathematics I (Herder)
16. SS 1762: Mathematics II

So we can conclude that Kant gave a two-semester course on *Auszug* for sixteen semesters, covering arithmetic, geometry, and trigonometry in the first semester and mechanics, hydrostatics, hydraulics, and aerometry in the second. Kant gave this course nine times, since he gave both parts of the course in the summer semester of 1757 and in the winter semester of 1759/60

18. [tr.] Arthur Warda, *Immanuel Kants Bücher.* Johann Friedrich Gensichen (1759–1807) inherited Kant's books. His books were auctioned off in April, 1808 after his death. Warda discovered the catalogue of this auction, *Verzeichniss der Bücher des verstorbenen Professor Johann Friedrich Gensichen, wozu auch die demselben zugefallene Bücher des Professor Kant gehören*, and so was able to find out what books Kant had owned at the time of his death. Warda lists the books with all the bibliographical information given in the catalogue. Martin does not give all this information but only what is required for his argument. In addition to full names of authors and more complete titles, I have also added the original titles of translated works where possible. (These are not in Warda.)

19. [tr.] This work deals with mathematics and its applications such as

arithmetic, geometry, geography, surveying, astronomy, navigation, mechanics, and theatrical effects; and civil, naval, and military engineering.

20. [tr.] The original catalogue published by Warda does not include an author. Presumably, the works are from the *Philosophical Transactions* of the Royal Society of London, the *Mémoires* of the Académie des Sciences, Paris, and others.

1. The Axiomatics and Logic of Mathematics

1. [tr.] The name *constructivism*, which is sometimes used for Brouwer's theory, is, I think, a better one than *intuitionism*. See my remarks about translating *Anschauung*, p. xiii.

2. [tr.] David Hilbert, *Grundlage der Geometrie*, 1. "So fängt denn alle menschliche Erkenntnis mit Anschauungen an, geht von da zu Begriffen und endigt mit Ideen." (Thus all human knowledge starts with perceptions, then goes on to concepts, and ends with Ideas.) Kant, CPR, A702 = B730.

3. [tr.] See, for example, Henri Poincaré, *Science et l'Hypothèse*, ch. 1.

4. Christian Wolff, *Elementa matheseos universae*, vol. 1, 29, 98.

5. [tr.] T. L. Heath raises the question of whether Euclid used *axiom* (*axioma*), which had already been used by Aristotle (384–322 B.C.) in discussions of mathematics, or *common notions* (*koinai ennoiai*), and concludes that Euclid used the latter term. *Euclid's Elements*, trans. Heath, vol. I, bk. I, 221–22.

6. [tr.] Imre Tóth argues that Aristotle himself thought of geometry as axiomatic in the sense given here by Martin. See his discussion of Aristotle's views about the parallel postulate in "Non-Euclidean Geometry before Euclid," *Scientific American*, 221 (1969), 87–92; also Tóth, "Das Parallelenproblem im Corpus Aristotelicum," *Archive for the History of the Exact Sciences*, vol. 3 (1966–67), nos. 4–5, 17 March 1967, 249–422.

7. [tr.] There seem to be only eight possibilities here, composed of combinations of one from each of the following pairs: arithmetic-geometry, logicist-axiomatic, deductive-constructive. Later, Martin discusses another pair, the analytic-synthetic. He may have this pair in mind when he says that there are sixteen possibilities.

8. Leibniz, "In Euclidis Prota," *Mathematische Schriften*, ed. C. I. Gerhardt, vol. V, 183–219.

9. "Animadversiones in partem generalem Principiorum Cartesianorum," ibid., vol. IV, 355.

10. Wolff, *Elementa matheseos universae*, De methodo mathematica, sec. 33.

11. Johann August Eberhard, "Ueber die apodiktische Gewisheit," *Philosophisches Magazin*, 2 (1789–90), 157.

12. Johann Christoph Schwab, *Commentatio in primum elementorum Euclidis librum, qua veritatem geometriae principiis ontologicis niti evincitur, omnesque propositiones, axiomatum geometricorum loco habitae, demonstrantur*, 44.

13. Hilbert, "Axiomatisches Denken," *Mathematisches Annalen* 78 (1908), 406.

14. Gauss, letter to Bessell, 9 April 1830, Ak. VIII, 201. [tr.] Friedrich Wilhelm Bessell (1784–1846).

15. Gottlob Frege, *Die Grundlagen der Arithmetik*, 101.

16. Vloemans, *Anschauung und Verstand*, 153.

17. Christian Gottlieb Zimmermann, *Anfangsgründe der Geometrie*, v.

2. The Analytic Principles

1. [tr.] Martin's text erroneously says "Axiome der Geometrie." The Kantian text in the second edition is "Axiomen der Anschauung." CPR, B202. (In the first edition the title is "Von den Axiomen der Anschauung." CPR, 162.)

2. Kant, CPR, A163ff. = B204ff.

3. Kant, *Prolegomena to Any Future Metaphysics*, 269; CPR, B16ff.

4. Johann Schultz, *Prüfung der Kantischen Kritik der reinen Vernunft*, I, 218ff. (See Appendix.)

5. [tr.] Martin here uses the Latin translations of the Greek terms that Euclid used, *koinai ennoia* (*common notions*, Latin *conceptiones communes*) and *aitémata* (*postulate*, Latin *postulata*). Cf. Euclid, *Elementa*, ed. E. S. Stamatis.

6. [tr.] The Greek for *elements* is *stoicheia*. Proclus discusses the meaning of *elements* in his *In primum Euclidis elementorum librum commentarii*. See the edition edited by G. Friedlein, 72 sqq.; Proclus says that elements are certain leading theorems in geometry which have to those which follow the relation of an all-pervading principle and which furnish proofs of many properties. Cf. Euclid's *Elements*, trans. T. L. Heath, I, chap. IX, secs. 1–2, 114–24.

7. Euclid, *Euclidis elementorum libri xv*, ed. C. Clavius.

8. Euclid, *Euclides restitutus*, ed. J. A. Borelli.

9. Henry Savile, *Praelectiones tresdecim in principium Elementorum Euclidis*; Euclid, *Euclidis elementorum libri xv*, ed. Isaac Barrow; Giovanni Girolamo Saccheri, *Euclides ab omni naevo vindicatus sive conatus geometricus quo stabiliuntur geometria principiae*.

10. Leibniz, *New Essays Concerning Human Understanding*, I, chap. III, sec. 24, Ak. 107–8. [tr.] Martin uses a German translation by Artur Buchenau, *Neue Abhandlungen über den menschlichen Vernunft*, ed. E. Cassirer. I use the Remnant and Bennett translation with minor changes. However, I translate the title differently, since the work Leibniz is commenting on is Locke's *Essay Concerning Human Understanding*. Leibniz's original title is *Nouveaux Essais sur l'Entendement Humain*. Theophilus speaks.

11. Leibniz, *New Essays*, trans. Remnant and Bennett (with minor changes), I, chap. II, secs. 24–27, Ak. 101. Theophilus speaks.

12. Johann Jakob Hentsch, *Philosophia mathematica complectens methodum cogitandi, nec non scientiam rerum universalem ex Euclide restitutam. Conamina duo priora*. [tr.] Martin identifies Hentsch as *E. Hensch*. Hentsch is also identified by the Latin version of his name, *Joannes Jacobus Hentschius*.

13. Georg Sarganeck, *Die Geometrie in Tabellen*.

14. Wenceslaus Joann Gustav Karsten, author of *Mathesis theoretica elementaris atque sublimior* and *Mathematische Abhandlungen*.

15. Georg Jonathan Holland, *Abhandlung über die Mathematik*, 38.

16. Jakob Friedrich Fries, *Die mathematische Naturphilosophie*, 61, 81.

17. Friedrich Schmeisser, *Lehrbuch der reinen Mathesis*, 130ff.

18. Alois (Aloys) Mayr, *Untersuchung über die wissenschaftliche Methode*, 143, 147, 150, 151.

19. Giuseppe Peano, *Arithmetices Principia*.

20. Leibniz, *Opuscules et fragments inedits*, ed. L. Couturat, 546.

21. Ibid., 255. Cf. the *Specimen Calculi universalis*, in *Philosophische Schriften*, ed. C. J. Gerhardt, vol. VII, 219.

22. Ibid.

23. Christian Wolff, *Philosophia prima, sive Ontologia*, sec. 181, Identitatis definitio (Definition of Identity).

24. Ibid., sec. 349, Aequalitatis et inaequalitatis definitio (Definition of Equality and Inequality), 274.

25. Schultz, *Anfangsründe der reinen Mathesis*, 29.

26. H. G. Grassmann, *Die Wissenschaft der extensiven Grössen oder die Ausdehnungslehre*, in *Gesammelte mathematische und physikalische Werke*, vol. I, 34.

27. Wolff, *Elementa matheseos universae*, vol. 1, 29.

28. Johann Heinrich (Jean Henri) Lambert, *Anlage zur Architektonik, oder Theorie des Einfachen und Ersten in der philosophischen und mathematischen Erkenntnis*, vol. I, 97. [tr.] Martin probably means sections 137–38, 97–99.

29. Schultz, *Anfangsgründe der reinen Mathesis,* 30.

30. [tr.] Proclus, *In primum Euclidis elementorum librum commentarii*, 183. T. L. Heath, in his *A History of Greek Mathematics*, vol. I, 192–93, makes the conjecture that Apollonius did this in a lost general treatise on mathematics which Marinus refers to in a work on Euclid's *Data*.

31. Savile, *Praelectiones tresdecim in principium Elementorum Euclidis*, 145.

32. Leibniz, "In Euclidis Prota," *Mathematische Schriften*, ed. Gerhardt, vol. V, 207.

33. Wolff, *Elementa matheseos universae*, 26; Wolff, *Philosophia prima, sive Ontologia*, secs. 366–67; Ibid., sec. 223.

34. Schultz, *Anfangsgründe der reinen Mathesis,* 30.

35. [tr.] Cf. Hermann von Helmholtz, "Zählen und Messen, erkenntnistheoretisch betrachtet."

36. Federigo Enriques, *Zur Geschichte der Logik*, 119.

37. Leibniz, *Opuscules*, 147.

38. Leibniz, "In Euclidis Prota," in *Mathematische Schriften*, ed. Gerhardt, vol. V, 206.

39. Schultz, *Anfangsgründe der reinen Mathesis*, 32.

40. Peano, *Arithmetices Principia*.

41. John Locke, *An Essay Concerning Human Understanding*, bk. IV, chap. vii, sec. 6. [tr.] Martin uses a German translation by Carl Winckler, *Versuch über den menschlichen Verstand*.

42. Wolff, *Philosophia prima, sive Ontologia*, sec. 341.

43. Ibid., sec. 352.

44. Ibid., sec. 356, p. 278.

45. Ibid., sec. 357, p. 278.

46. Leibniz, "Initia rerum mathematicarum metaphysica," in *Mathematische Schriften*, ed. Gerhardt, vol. VII, 20.

47. Leibniz, *New Essays*, trans. Remnant and Bennett, bk. I, chap. III, sec. 6, Ak. 102–3.

48. Schultz, *Anfangsgründe der reinen Mathesis*, 38.

49. Giulio Vivanti, "Infinitesimalrechnung," in *Vorlesungen über Geschichte der Mathematik*, ed. Moritz Cantor, vol. 4, pt. XXVI, 658.

50. Louis Couturat, "La philosophie des mathématiques de Kant," in *Revue de Métaphysique et de Morale*, 12 (1904), 346.

51. Christian August Crusius, *Weg zur Gewissheit und Zuverlässigkeit der menschlichen Erkenntnis*, 470.

52. Kant, *Enquiry into the Clarity of the Principles of Natural Theology and Morals*, Third Reflection, sec. 3, Ak. II, 293–96.

53. Ibid., 295.

54. Crusius, *Weg zur Gewissheit und Zuverlässigkeit der menschlichen Erkenntnis*, 473.

55. Kant, *Principles of Natural Theology and Morals*, Ak. II, 295ff, 282, 281.

56. Ibid.

57. Ibid.

58. J. S. Beck, letter to Kant, 24 August 1793, Ak. XI, 444ff. [tr.] Karl Christian Erhard Schmid(t) (1761–1812).

59. Kant, *On the Forms and Principles of the Sensible and Intelligible, Worlds*, Ak. II, 395.

60. [tr.] *Intuitus*. See my earlier remarks on translating *Anschauung*.

61. Kant, *On the Sensible and Intelligible Worlds*, Ak. II, 397.

62. Ibid., 410.

63. Konrad Dieterich, *Kant und Newton*, 109.

64. Kant, *Streitschrift gegen Eberhard*, Ak. VIII, 196. [tr.] This is the essay Kant wrote in 1790 in answer to Johann August Eberhard's attacks, *On A Discovery According To Which All New Critiques Of Pure Reason Are Dispensable Because of An Older One* (*Über eine Entdeckung, nach der all neue Kritik der reinen Vernunft durch eine ältere entbehrlich gemacht werden soll*) See Heinrich Maier's introduction to the essay in Ak. VIII, 492–495. See also Henry E. Allison's *The Kant-Eberhard Controversy*, which also has an English translation of Kant's essay.

65. Kant, Ref. 3709, Ak. XVII. [tr.] The Reflexions in this volume are classified by the editors as those of metaphysics. The volume includes Reflexions 3489–4846.

66. Ibid., Ref. 3747.

67. Ibid., Ref. 3710.

68. Ibid., Ref. 4655.

69. Ibid., Ref. 3742.

70. Ibid., Ref. 3923.

71. Ibid., Ref. 3750.

72. Ibid., Ref. 3744.

73. Ibid., Ref. 3976.

74. Kant, Ref. 3127, Ak. XVI. [tr.] The Reflexions in this volume are classified by the editors as those of logic. The volume includes Reflexions 1526–3488.

75. Ibid., Ref. 3126.

76. Kant, Ref. 3923, Ak. XVII.

77. Ibid., Ref. 4370.

78. Ibid., Ref. 4477.

79. Ibid., Ref. 4162.

80. Kant, letter to Marcus Herz, 21 February 1772, Ak. X, 131.

81. Johann Gottfried Karl Christian Kiesewetter, *Die ersten Anfangsgründe der reinen Mathesis*, sec. 9.

82. Christian Gottlieb Zimmermann, *Entwicklung analytischer Grundsätze für den ersten Unterricht in der Mathematik*, 4.

83. Ludwig Heinrich Jakob, *Grundriss der allgemeinen Logik und kritischen Anfangsgründe zu einer allgemeinen Metaphysik, chap.* 1, secs. 97–121, 83–96.

84. Georg Simon Klügel, *Mathematisches Wörterbuch*, vol. II, 696.

85. Hermann Hankel, *Theorie der komplexen Zahlensysteme*, 52.

86. Günther Thiele, W*ie sind die synthetischen Urteile der Mathematik a priori möglich?*, 4.

87. Euclid, *Die Elemente, Gesamtausgabe*, ed. J. L. Heiberg and H. Menge, trans. Clemens Thaer, 7 vols.; Euclid, *Euclidis elementorum libri xv,*, ed. C. Clavius.

3. The Axioms of Arithmetic

1. Johann Schultz, *Anfangsgründe der reinen Mathesis*, 32, 40, 41.

2. Leibniz, "In Euclidis Prota," *Mathematische Schriften*, ed. Gerhardt, vol. V, 206. [tr.] In the section entitled "Ad libri primi Euclidis postulata," Leibniz writes as Postulate IV: "With any given magnitude, it is possible to add a greater or a smaller." (*Quavis magnitudine data sumi posse majorem vel minorem*).

3. Leibniz, "Prima calculi magnitudinum elementa demonstrata in additione et subtractione, usuque pro ipsis signorum + et − ," *Mathematische Schriften*, ed. Gerhardt, vol. VII, 77–82.

4. Schultz, *Prüfung der Kantischen Critik der reinen Vernunft*, I, 221ff.

5. Christian Wolff, *Philosophia prima, siva Ontologia*, sec. 397.

6. Lambert, *Neues Organon, oder Gedanken über die Erforschung und Beziehung des Wahren und dessen Unterscheidung von Irrtum und Schein*, vol. I, 469. [tr.] Compare this with the account of the origin of the ideas of number in Descartes: "As for my clear and distinct ideas of corporeal things, there are some which, it seems to me, might have been taken from my ideas of myself, such as my ideas of substance, duration, number, and the like. . . . when I think I exist now, and recollect besides that I existed some time ago, and when I am conscious of various thoughts whose number I know, I then acquire the ideas of duration and number, which I can afterwards apply to anything else I please." René Descartes, *Meditations on First Philosophy*, trans. L. J. Lafleur (with minor changes), Med. III, Adam and Tannery edition Latin text 44–45, French text 35.

7. Lambert, *Neues Organon*, I, 500.

8. J. G. Kiesewetter, *Die ersten Anfangsgründe der reinen Mathematik*, 20, 23.

9. [tr.] Martin probably means Christian Gottlieb Zimmermann's *Entwicklung analytischer Grundsätze für den ersten Unterricht in der Mathematik* of 1805.

10. C. H. Müller gives information about Murhard in the note in *Studien zur Geschichte der Mathematik in Göttingen, Abhandlungen zur Geschichte der mathematischen Wissenschaften*, 18 (1904), 137.

11. Friedrich Wilhelm August Murhard, *System der Elemente der allgemeinen Grössenlehre*, Lemgo, 1798, 38. [tr.] Compare this with the passage above from Schultz's *Anfangsgründe*, 32, 40, 41, about the axioms and postulates of arithmetic. I have translated this so as to make clear the word for word correspondence. Differences in the German I have indicated by differences in the English.

12. Carl Friedrich Gauss, letter to Wolfgang Bolyai, 9 January 1799, *Briefwechsel*, eds. F. Schmidt and P. Stackel, 15. [tr.] Johann Andreas Segner (1704–77); Konrad Dietrich Martin Stahl (1771–1833), *Zahlenarithmetik und Buchstabenrechnung*, Jena, 1797. As Martin explains, "Schulze" is Johann Schultz. "Pfaff" is probably Johann Friedrich Pfaff (1765–1825), then professor of mathematics at the university in Helmstedt and a good friend of Gauss. See Gauss's letter to Bolyai, 16 December 1799, where he describes how he is living at Pfaff's home, *Briefwechsel*, 36.

14. [tr.] The text says "Anfangsgründe" but this seems to be incomplete, since Martin also shows the word for word correspondence with passages in Schultz's *Prüfung.*.

15. [tr.] German *sich erweiternde Wissenschaft*. The related *Erweiterungsurteile* in the *Critique of Pure Reason*, A7 = B11, is translated as *expanding judgments* by F. Max Müller and as *ampliative judgments* by Norman Kemp Smith.

16. [tr.] All three passages from Kant, Schultz, and Murhard respectively are from the same writings: Kant's letter to Schultz, 25 November 1788, 555; Schultz's *Prüfung*, 1798, 232; and Murhard's *Systeme der Elemente der allgemeinen Grössenlehre*, 1798, 39.

17. Friedrich Leopold Freiherr von Hardenberg (Novalis), *Schriften und Briefe*, vol. III, 23. [tr.] In notes of 1798 on Murhard's *System der Elemente der allgemeinen Grössenlehre*, Novalis writes: "Schultz has already supplied the elements for calculating infinite quantities in his *Attempt at a Precise Theory of the Infinite*." [*Schultz hat in dem Versuch Theorie des Unendl[ichen] - die Anf[angs] Gr[ünde] der Berechnung des Unendl[ich] Grossen bereits geliefert.*] 3d ed., vol. III, 120.

18. Martin Ohm, *Kritische Beleuchtungen der Mathematik*, 27ff.

19. Jakob Friedrich Fries, *Die mathematische Naturphilosophie*, 61, 81ff.

20. Ibid., 112, 258, 260, 263, 276, 277, 285, 286, passim.

21. Ohm, *Versuch eines vollkommen consequenten Systems der Mathematik*, pt. I, 14, 16.

22. Ibid., pt. I, 13.

23. Ohm, *Elementar-Zahlenlehre*, 7.

24. Ibid., 11.

25. Schultz, *Prüfung*, pt. I, 220.

26. [tr.] The *e* is for the German *Einheit*, which means *unit*.

27. H. G. Grassmann, *Lehrbuch der Arithmetik und Trigonometrie für Höhere Lehranstalten*, in *Gesammelte mathematische und physikalische Werke*, vol. II, pt. 1, sec. 2 "Addition," 300, 301, 303.

28. Grassman, *Die Wissenschaft der extensiven Grössen oder die Ausdehnungslehre*, in the *Werke*, vol. I, pt. 1, 22.

29. Hermann Hankel, *Theorie der komplexen Zahlensysteme*, 33.

30. Ibid., 40.

31. [tr.] The Archimedean axiom is the axiom that any magnitude can be taken enough times so that the resulting magnitude will be larger than a given assigned magnitude; in contrast to the principle underlying the calculus that sometimes this is not so, namely, sometimes infinitesimal magnitudes can be taken as much as one pleases and yet the resulting magnitude will not be larger than a given magnitude. E.g., the sum $1 + 1/2 + 1/4 + 1/8 + \ldots$ in which each succeeding number is half of the preceding one, as in Zeno's paradox of the man trying to walk out of a stadium, will never be greater than 2. See T. L. Heath on the history of this in his *Greek Mathematics*, 193, and his edition of Euclid's *Elements*, vol. I, 234 and vol. III, 15–16. The original proposition by Archimedes was actually a lemma stated in the preface to his *Quadrature of a Parabola*: "The excess by which the greater of two unequal areas exceeds the less can, if it be continually added to itself, be made to exceed any assigned finite area." T. L. Heath, *Euclid's Elements*, vol. I, 234. See also Heath's discussion in vol. III, 15–16 and in his *Greek Mathematics*, 193–4.

32. Gottlob Frege, *The Foundations of Arithmetic*, 8.

33. Giuseppe Peano, *Arithmetices Principia*.

34. [tr.] Martin presumably is referring to Mortiz Cantor. I have been unable to locate Martin's source.

35. Schultz, *Anfangsgründe der reinen Mathesis*, 64.

36. Ohm, *Versuch eines vollkommen consequenten Systems der Mathematik*, pt. I, 57ff.

37. Grassmann, *Lehrbuch der Arithmetik*, in *Werke*, vol. II, pt. 1, 321.

38. Sir William Rowan Hamilton, *Lectures on Quaternions*, 3, fn.

39. [tr.] Martin's scrupulous caution was well-justified. Thomas L. Hankins, in his recent biography of Hamilton, argues that Hamilton came to his view that algebra was the science of pure time before he read the *Critique of Pure Reason. Sir William Rowan Hamilton*, 258; cf. chaps. 17–19.

40. Hamilton, *Lectures on Quaternions*, 18, second fn. [tr.] This is as Hamilton writes it, not as Martin does.

41. Peano, *Arithmetices Principia*, v.

42. H. Hankel, *Theorie der komplexen Zahlensysteme*, 15.

43. Friedrich Engel, "Grassmanns Leben," in Grassmann, *Gesammelte Werke*, vol. III, pt. 2, 75, fn. 1.

44. [tr.] The following is according to Gerhard Lehmann's introduction in the appendix of Ak. XX, 483–488: Wilhelm Dilthey, "Neue Kanthandschriften,"

Kant-Studien, 3 (1899). Abraham Kästner wrote three articles for the magazine of Johann August Eberhard, the *Philosophisches Magazin*, 2 (1790): 1) "Was heisst in *Euklids* Geometrie möglich??" (391–402), 2) "Ueber den mathematischen Begriff des Raums" (403–19), 3) "Ueber die geometrischen Axiome" (420–30). Johann Schultz wrote a review of these articles for the *Jenaer Litteraturzeitung* in August 1790. This was all part of the dispute between Kant and Eberhard, and Kant sent material for Schultz to include in the review. See Ak. XX, 410–23 and Kant's letters to Schultz of June 1790, 29 June 1790, and 2 August 1790, Ak. XI, 182–84.

45. Schultz, *Erläuterungen über des Herrn Professor Kant Critik der reinen Vernunft*. Cf. letter from Schultz to Kant, 28 August 1783, Ak. X, 352ff. and letter from Kant to Schultz, 17 February 1784, Ak. X, 366ff. [tr.] I am unclear why Martin says that Kant took no direct interest in this work. In his letter Kant criticizes the argument in Schultz's manuscript that only two categories are necessary, since the third follows from the first two; and he explains at length why.

46. [tr.] The title of Schultz's inaugural dissertation of 15 February 1787 at the University of Königsberg was D*e geometrica acustica nec non de ratione 0:0 seu basi calculi differentialis.*

47. Kant, letter to Schultz, 25 November 1788, Ak. X, 554–58.

48. Ibid., 555–56.

49. Jacob Sigismund Beck, *De theoremate Tayloriano, sive de lege generali*, Thesis 5.

50. Schultz, *Prüfung*, pt. I, 217.

51. Ibid., 235.

52. [tr.] Martin should also include the commutative law of multiplication, since he uses it in this chapter as an example.

4. Problems about Classes of Numbers

1. [tr.] Abraham Gotthelf Kästner, *Anfangsgründe der Arithmetik, Geometrie, ebenen und sphärischen Trigonometrie und Perspective*, Göttingen, 1758. In the third ed., 1774, Kästner defines negative quantities as follows: "Opposed quantities are called quantities such that, when under this condition, the one is regarded to decrease the other," 62. Kant refers to this work in *An Attempt to Introduce the Concept of Negative Quantities into Philosophy*, Ak. II, 170. See Kurd Lasswitz's note on this passage, Ak. II, 479. It is unclear why Martin says that Kant showed his greatness by being aloof from questions like that about negative numbers, since Kant wrote about negative numbers in his essay, *An Attempt to Introduce the Concept of Negative Quantities into Philosophy*.

2. Wenceslaus Joann Gustav Karsten, *Mathematische Abhandlungen*.

3. Johann Schultz, *Anfangsgründe der reinen Mathesis*, 131.

4. Kant, *An Attempt to Introduce the Concept of Negative Quantities into Philosophy*, Ak. II, 174.

5. Schultz, *Anfangsgründe*, 122. [tr.] Martin adds the following remark: Note that there is nothing in Kant which suggests Hankel's idea of the permanence of

the formal laws. Hence, the Marburg school interpretation of synthesis, which is mainly from this point of view, is hardly relevant.

6. Leonhard Euler, *Vollständige Anleitung zur Algebra*, 2 vols., Petersburg, 1770; in *Opera omnia*, vol. I, para. 14, 13. [tr.] I have added paragraph 13 for reference. "The formula above" refers to paragraph 13 immediately preceding, 12–13.

7. Ibid., para. 259, 93.

8. Carl Hadaly von Hada, *Anfangsgründe der Mathematik*, 23.

9. Kant, *The Concept of Negative Quantities*, Ak. II, 170.

10. [tr.] Morris Kline makes a similar point, arguing that mathematics without a basis in the empirical world is sterile. *Mathematics: The Loss of Certainty*.

11. Kant, Ref. 2885, Ak. XVI, 559–560. [tr.] This volume of Reflexions about logic consists of Kant's comments about George Friedrich Meier's *Auszug aus der Vernunftlehre* (Halle, 1752). This is an extract from Meier's *Vernunftlehre*, which appeared at the same time. Martin cites the second edition of *Vernunftlehre* (Halle, 1762). Ak. XVI contains both the *Auszug* and Kant's corresponding comments. Martin makes the following remark:

> Compare [Ref. 2885] with what Meier says in the *Vernunftlehre*: "Briefly, separation is, in fact, a subtraction by which we take away the differences from lower concepts until we get the very highest concept." *Vernunftlehre*, sec. 294, pp. 430ff. This is quoted in Ak. XVI, 560, ll. 23–25. The same comparison is found somewhat more developed in Reflexion 5652; however, nothing of any particular significance can be derived even from this Reflexion.

12. Kant, letter to August Wilhelm Rehberg, 25 September 1790, Ak. XI, 196ff.

13. René Descartes, *Geometria*, I, 76. [tr.] Cf. *La Geometrie*, in the *Oeuvres*, eds. C. Adam and P. Tannery, vol. 6, Livre III, 380. Also see *The Geometry*, trans. by David Eugene Smith and Marcia L. Latham, pagination of the French Adam and Tannery edition (the 1637 edition), and *Discourse on Method, Optics, Geometry, & Meteorology*, trans. Paul J. Olscamp, 236. For an eighteenth century account of the history and meaning of imaginary numbers, see Charles Hutton, *Mathematical & Philosophical Dictionary*, London, 1795, article on Imaginary Quantities, 625–7.

14. [tr.] Colin Maclaurin, *A Treatise of Algebra*, part II, chap. 1, secs. 8–10, 131–296.

15. Leibniz, *Specimen novum analyseos*, in *Mathematische Schriften*, ed. Gerhardt, vol. V, 357.

16. Schultz, *Anfangsgründe*, 170.

17. [tr.] Although William Jones introduced in 1706 the use of π to designate the ratio of the circumference of a circle to its diameter (*Synopsis palmariorum matheseos*, 263), it only came into wider use after Euler used it in his *Introductio in analysin infinitorum* in 1748. Cf. H. C. Schepler, "The Chronology of Pi," *Mathematics Magazine*, 1950, 223, 224; F. Cajori, *History of Mathematical Notations*, Vol. II, 9. For a lively general history, see P. Beckmann, *A History of π (Pi)*.

18. Schultz, *Sehr leichte und kurze Entwicklung einiger der wichtigsten mathematischen Theorien*, Königsberg, 1803.

19. Lambert, "Mémoires sur quelques propriétés remarquables des quantités transcendentes circulaires et logarithmique," in *Histoire de l'Academie Royale des Sciences et des Belles Lettres de Berlin*, 17 (1768), 265–324.

20. [tr.] Roland Häggkvist has explained the memoir to me as follows. First, *transcendent* quantities should not be confused with *transcendental* numbers. Transcendental numbers are numbers which are not the roots of algebraic equations with rational coefficients. By *transcendent*, Lambert means quantities which *transcend* from a circle, like sines, cosines, and π. Second, Lambert proves in this memoir that π is irrational but not that it is transcendental. In sections 89–90 he makes the conjecture that the quantities he has proved to be irrational, π in particular, are not the roots of any algebraic equations (with integer coefficients), noting that, for example, if the cos 2ω is irrational, then the tan ω^2 is not the square root of a rational number. C. L. F. (Ferdinand) Lindemann (1852–1939) finally proved in 1882 that π was transcendental. Ferdinand Lindemann, "Ueber die Zahl Pi," *Mathematische Annalen*, 20 (1882), 213–225. See also D. E. Smith, "The History and Transcendence of Pi," in *Monographs on Modern Mathematics*, 387–416.

21. Anton Edler von Braunmühl, "Trigonometrie, Polygonometrie und Tafeln," *Vorlesungen über die Geschichte der Mathematik*, ed. M. Cantor, vol. IV, pt. XXIII, 448.

22. Christian Wolff, *Philosophia prima, siva Ontologia*, sec. 405, p. 312.

23. Kant, CPR, A480 = B508.

5. *Combinatorics and the Idea of a Systematic Ontology*

1. [tr.] I have not been able to locate Martin's source. Luca Paccioli, also known as Luca Pacioli or Luca de Burgo (ca. 1445–1509) explains how to calculate the number of possible combinations of people sitting around a table for 1 person, 2 persons, up to any number of persons, *n*, in his *Summa de Arithmetica, geometria, proportioni: et proportionalita*, 1st ed. 1494; ed. of 1523, folio 43 verso. For a discussion of combinatorics in antiquity see Adolphe Rome, "Procédés anciens de calcul des combinaisons," *Annales de la Société scientifiques de Bruxelles*, series A, 50 (1930), 97–104.

2. [tr.] Cf. Blaise Pascal, "Traité du Triangle Arithmétique," sec. II; "Combinationes"; and "Des Combinaisons"; in *Oeuvres Complètes*, 55–57, 76–83. Also see Jacob Bernoulli, *Ars conjectandi* and Leibniz's *De Arte Combinatoria*.

3. [tr.] S. (A. W. Siegmund) Günther (1848–1923) calls Karl Friedrich Hindenburg the "Begründer der kombinatorischen Analysis" (founder of combinatorial analysis) in his "Geschichte der Mathematik," in *Vorlesungen über Geschichte der Mathematik*, ed. M. Cantor, vol. 4, 4–5.

4. [tr.] Martin adds: For pure mathematics, consult Gauss's contemporary, Bernard F. Thibaut, in his time considered the best Dozent of mathematics, who wrote numerous widely read textbooks; for the problems of foundations, see Lambert and Fries.

5. [tr.] See for example Ramón Lull (Raymundus Lullus), *Ars Generalis Ultima* and *Ars Brevis*. Martin Gardner has a delightful account of Lull's *Ars Magna* in his *Logic Machines, Diagrams and Boolean Algebra*, chap. 1.

6. [tr.] Latin *logica inventiva*. Logic was traditionally divided into *invention* or *inventive logic* or the *logic of discovery* and the *logic of judgment* or the *analytic part of logic*, a division which stemmed from Cicero. Cf. Cicero, *Topics*, ii, 6 and Boethius, *Second Commentary on the Isagoge of Porphyry*, I, 2; G. H. R. Parkinson, "Introduction to Leibniz," *Logical Papers*, xii including fn. 4; Wilhelm Risse, *Die Logik der Neuzeit*, vol. I, 18–19.

7. Leibniz, in *De Arte combinatoria*, in *Philosophische Schriften*, vol. IV, 61.

8. Ibid., 65.

9. For details see Louis Couturat, *La Logique de Leibniz d'après des documents inédits*, in particular chaps. 4, 6, 8.

10. Leibniz, *Philosophische Schriften*, vol. VII, 300; ibid., vol. VIII, 6; Louis Couturat, *La Logique de Leibniz*, 208.

11. Ibid.

12. Christian Wolff, *Philosophia rationalis, sive Logica*, sec. 61.

13. Ibid., sec. 223.

14. Ibid., sec. 234.

15. Herman Schmalenbach, *Leibniz*, 502 and chap. VII.

16. Ibid., 141.

17. Lambert, letter to G. J. Holland, 7 April 1766, in Lambert, *Deutscher gelehrter Briefwechsel*, vol. I, 136.

18. Kant, Ref. 5024, Ak. XVIII.

19. Ibid., Ref. 4866.

20. Kant, Ref. 1629, Ak. XVI.; Refs. 4893, 4900, Ak. XVIII.

21. Kant, CPR, A65 = B90.

22. Kant, CPR, A66ff. = B91ff.

23. Lambert, *Neues Organon, oder Gedanken über die Erforschung und Bezeichnung des Wahren und dessen Unterscheidung von Irrtum und Schein*, and *Anlage der Architektonik, oder Theorie des Einfachen und Ersten in der philosophischen und mathematischen Erkenntnis*.

24. Lambert, *Neues Organon*, vol. I, 456.

25. Ibid., 421.

26. Lambert, *Architektonik*, vol. I, 6.

27. Ibid., 30.

28. Ibid., 12ff.

29. Lambert, *Neues Organon*, vol. I, 498.

30. Ibid., 527.

31. Ibid., 565.

32. Ibid., 538.

33. Kant, *Enquiry into the Clarity of the Principles of Natural Theology and Morals*, First Reflection, sec. 2, Ak. II, 279.

34. Kant, *A New Exposition of the First Principles of Metaphysical Knowledge*, Ak. I, 389f. [tr.] Hermann Boerhaave (1668–1738), *Elementa chimiae*, vol. I, 65. Cf. Kurd Lasswitz's note on the passage, Ak. I, 566. Aesop (619–546 B.C.), "The Farmer and His Sons," *Fables,* trans. V. S. Vernon. See also Ben Edwin Perry, *Aesopica,* Pars Quinta, Fabulae Graecae, 42, p. 338. Francis Bacon (1561–1626) also tells this fable in describing how alchemists may have made many discoveries despite their foolishness. *Novum Organum,* I, Ap-

horism 85. The motto to the second edition of the *Critique of Pure Reason* from the *Instauratio Magna* shows that Kant had certainly read the *Novum Organum*.

35. Kant, *The Basic Reason for the Distinction of Regions of Space*, Ak. II, 377. [tr.] H. Boerhaave, *Elementa chemiae*, vol. I, 2. Cf. Kurd Lasswitz's note, Ak. II, 508.

36. Kant, CPR, The Discipline of Pure Reason, sec. I: The Discipline of Pure Reason in Its Dogmatic Employment, A712ff. = B740ff.

37. Kant, CPR, A724 = B752.

38. Kant, Ref. 4937, Ak. XVIII.

39. Ibid., Ref. 4938.

40. Ibid., Ref. 5047.

41. Kant, letter to Marcus Herz, end of 1773, Ak. X, 144.

42. Kant, letter to Schultz, 26 August 1783, Ak. X, 351.

43. Schultz, letter to Kant, 28 August 1783, Ak. X, 354.

44. Kant, letter to Jacob S. Beck, 27 September 1791, Ak. XI, 290.

45. Ref. 1946, Ak. XVI.

46. Kant, Ref. 3711, Ak. XVII. Cf. Refs. 3873, 3989, 3997, 4391, 4658, 5651.

47. Kant, *Thoughts about the True Evaluation of Living Forces*, Ak. I, 10.

48. Kant, *The Concept of Negative Quantities*, Ak. II, 203ff.

49. Kant, *Dreams of a Spiritseer*, Ak. II, 342.

50. Kant, CPR, A80 = B106 (according to no. XLIV of the *Nachträge*). [tr.] Martin quotes this sentence as it is supposed to have been amended by Kant, deleting the word *ursprünglich* (original), which appears in the first and second editions. See footnote 2 to this sentence in the edition of Raymund Schmidt, *Philosophische Bibliothek* vol. 37a. See also the *Nachträge, Textemendationen* to the first edition, Ak. XXIII, 46, A80.

51. Ibid., A81 = B107; ibid. A64 = B89, A83 = B109; ibid., A81 = B107.

52. Ibid., A81 = B107.

53. Ibid., A82 = B108.

54. Ibid., AXXI.

55. Ibid., A856 = B884.

56. Ibid., B110, footnote.

57. Kant, letter to Moses Mendelssohn, 16 August 1783, Ak. X, 346.

58. Kant, letter to Johann Bering, 7 April 1786, Ak. X, 441.

59. Kant, *Some Remarks on Ludwig Heinrich Jakob's Examination of Mendelssohn's Morgenstunde* [*Einige Bemerkungen zu Ludwig Heinrich Jakobs Prüfung der Mendelssohnschen Morgenstunde*], intro. essay to Ludwig Heinrich Jakob, *Prüfüng der Mendelssohnschen Morgenstunde . . . Nebst einer Abhandlung von Herrn Professor Kant*, p. LX]; Ak. VIII, 155.

60. Kant, letter to Herz, 24 December 1787, Ak. X, 512.

61. Kant, letter to Beck, 20 January 1792, Ak. XI,, 313ff.

62. [tr.] Kant, *Über die von der Königl. Akademie der Wissenschaften zu Berlin für das Jahr 1791 ausgesetzte Preisfrage*: "Welches sind die wirklichen Fortschritte, die die Metaphysik seit Leibnitzens und Wolffen's Zeiten in Deutschland gemacht hat?" (On the Prize Question Set by the Berlin Royal Academy of Science for 1791: What Are the Actual Advances Metaphysics Has Made in Germany Since the Time of Leibniz and Wolff?), *Sämtliche Werke*, ed.

Karl Vorländer, vol. 5; *Kleinere Schriften zur Logik und Metaphysik*, ed. Karl Vorländer, pt. 3, *Philosophische Bibliothek* series, 46c. Right after Kant's death in 1804, Friedrich Theodor Rink edited and published parts of drafts Kant had written on this topic; Kant never published a finished version. In the Academy edition it is in vol. XX, (*Handschriftlicher Nachlass*, vol. VII), 253–351. See Rink's Foreword, Ak. XX, 257–258, and Gerhard Lehmann's Introduction, Ak. XX, 479–483. The numbers in parentheses in the chart are the pages in the Academy edition. See the invaluable *Kant-Seitenkonkordanz* by Norbert Hinske and Wilhelm Weischedel, which is a concordance of the pagination of Kant's works in the different editions.

63. Kant, CPR, A81 = B107.

64. Kant, *Advances in Metaphysics*, in *Zur Logik und Metaphysic*, ed. K. Vorländer, *Philosophische Bibliothek*, vol 46c, 98; Ak. XX, 272–73.

65. Ibid., Vorländer ed., 84; Ak. XX, 260.

66. Kant, CPR, A841 = B869.

67. Paul Natorp, *Die Logische Grundlagen der exacten Wissenschaften*, 276.

68. Kant, *Enquiry into the Clarity of the Principles of Natural Theology and Morals*, Second Reflection, Example, Ak. II, 290.

69. [tr.] Evert W. Beth (1908–1964) argues that Kant's ideas about analytic and synthetic methods are like those of a Dutch thinker, Bernard Nieuwentyt (1654–1718), who distinguished between the ideal method of pure mathematics (synthetic) and the real method of applied mathematics (analytic). *The Foundations of Mathematics*, 44. See also Beth, "Nieuwentyt's Significance for the Philosophy of Science," *Synthese*, 9 (1953–55), 447–453, where he discusses Nieuwentyt's *Gronden van Zekerheid af de regte betoogwyze der Wiskundigen so in het denkbeeldige als in het zkelijke: ter weeklegging van Spinosa denkbeeldig samenstil in ter aanleidung van eene sekere sakelyke wysbegeerte, aangetoont door* (Fundaments of Certitude, or the Right Method of Mathematicians in the Imaginary As Well As in the Real: demonstrated in order to refute Spinoza's Imaginary System: and to introduce a Certain Real Philosophy). Amsterdam, 1720; 2d ed., 1728; 3d ed., 1739; 4th ed., 1754).

Kant was familiar with Nieuwentyt for he refers to him in *Der einzig mögliche Beweisgrund zu einer Demonstration des Daseins Gottes*, (The Only Possible Way of Proving the Existence of God), 1762, Ak. II, 160. Paul Menzer's note on Ak. II, 473 only mentions Nieuwentyt's other book, *Het regt Gebruik der Weltbeschouningen, ter overtuiginge van ongodisten en ongelovigen aangetoont* (The Right Use of Contemplating The Works of the Creator), Amsterdam, 1716, which went through at least seven printings in the Netherlands and was translated into English, French, and German (Frankfurt and Leipzig, 1732; Jena, 1747). The *Gronden* went through at least four editions in the Netherlands, which indicates its popularity there. However, thus far I have been unable to locate any German translations.

The distinction between analytic and synthetic methods in mathematics in fact goes back to the Greeks. See T. L. Heath's discussion in his edition of Euclid's *Elements*, vol. I, 138–42. Definitions of them are interpolated in Euclid XIII, 1–5; see Heath's comment on XIII, 1. Pappus has a fuller account. See *Pappi Alexandrinus Collectionis*, bk. VII, 1–2, pp. 634–717. Or see *Der Sammlung des Pappus von Alexandrien*, bk. VII, pp. 2–5. James Gow also has an excellent discussion of this in his *A Short History of Greek Mathematics*, sec. 108, pp. 177–180. Jaako Hintikka discusses this in his "Kant and the Tradition of Analysis," in *Deskription, Existenz und Analytizität*, 254–71. See also Hintikka and Unto Remes's *Method of Analysis*. Jürgen Mittelstrass in an unpublished paper given at the Leibniz conference in Toronto, November 1982, "Leibniz and Kant on Mathematical and Philosophical Knowledge," discusses the meaning of these terms in the seventeenth and eighteenth centuries. He holds that Kant was not familiar with the history of these terms at the time he wrote the *Enquiry into the Clarity of the Principles of Natural Theology and Morals*, though he was by the time he wrote the *Critique of Pure Reason*.

70. Kant, CPR, A13f. = B27f.

71. See above, pp. 68f.

72. Kant, *Metaphysical Foundations of Natural Science*, Ak. IV, 478.

73. Ibid., 473. Cf. *Advances in Metaphysics*, in *Zur Logik und Metaphysik*, ed. Vorländer, 98; Ak. XX, 272–73.

74. Kant, *Advances in Metaphysics*, in *Zur Logik und Metaphysik*, ed. Vorländer, 87; Ak. XX, 262.

75. Kant, *Advances in Metaphysics,* 88; Ak. XX, 263.

76. Kant, CPR, A713 = B741.

77. Kant, *Metaphysical Foundations of Natural Science*, Ak. IV, 469.

78. [tr.] Martin is using Johann Schultz's terminology here. See Schultz's *Prüfung der Kantischen Critik der reinen Vernunft*, 211:

> Die *Grösse* oder *Quantität* eines Dinges heisst diejenige innere Bestimmung derselben, die durch die Verbindung des Gleichartigen erzeugt wird. Das Ding selbst, das eine Quantität hat, heisst eine Grösse *in concreto*, oder ein *Quantum*.
>
> [The *magnitude* or *quantity* of a thing is its inner determination which is generated from the combining of the homogeneous. The thing itself that has quantity is called a quantity *in concreto* or a *quantum*.]

See the Appendix for the full text.

79. Kant, CPR, B49. [tr.] C. T. Michaelis goes into detail about the relation between time and arithmetic for Kant in his "Über Kants Zahlbegriff," *Wissenschaftliche Beilage zum Programm der Charlottenschule*. Berlin, 1884. He argues that Kant was uncertain about it.

80. Kant, CPR, B41.

81. [tr.] Martin seems to refer here to Schultz's *Prüfung,* 236. If so, he misinterprets him because Schultz does indeed think there is a special science of time, pure mechanics:

> Allein man hat heibey nicht erwogen, dass wir in der That eine *ganze reine Zeitwissenschaft* haben, die eben nicht von zu kleinem Umfange ist, nemlich die *reine Mechanik*, deren Hauptsache vorzüglich Zeitmessung ist, und es ist vielmehr zu verwundern, dass die Wissenschaft, die die Zeit zum Objecte hat, noch wirklich von so grossem Umfange ist, indem die Zeit nicht nur bloss eine einzige Dimension hat, sondern auch bloss einer *geraden* Linie correspondiert, und überdem nur eine *einzige* unendliche gerade Linie vorstellt.
>
> [This, however, overlooks that in point of fact we do have an *entire pure science of time* which is of no small scope, namely, *pure mechanics*, whose main objective is, above all, the measurement of time. It is more a cause of wonder that the science

which has time as its object really has so great a scope, since time has only a single dimension and also merely corresponds to a *straight* line, and what is more, only to a *single*, infinite straight line.]

Schultz takes credit for the discovery of the axioms of time. He describes pure mechanics as the pure science of time to answer Kant's critics who ask why the perceptual nature of time has not helped us discover any theorems, although that of space has helped us get the whole science of geometry. It is unclear to me whether Schultz means to credit Kant with this view of pure mechanics, or whether he is just adding his own argument against the critics. See the Appendix for the full text.

82. Kant, CPR, A82 = B108.

6. Synthetic Judgment in Arithmetic

1. Kant, *On the Forms and Principles of the Sensible and the Intelligible Worlds*, Ak. II, 397f.

2. See Hans Vaihinger, *Kommentar zu Kants Kritik der reinen Vernunft*, pt. 2, 387. See also Oskar Becker, "Mathematical Existenz," *Jahrbuch für Philosophie und phänomenologische Forschung*, 7 (1927), 654.

3. Kant, CPR, A78 = B104.

4. Ibid., B111.

5. [tr.] See H. W. Fowler's note on *manifold*, which is not idiomatic English now. *Dictionary of Modern English Usage*, 351.

6. Kant, CPR, A142ff. = B182ff. [tr.] The Latin in parenthesis is Kant's. Compare his usage with that of Schultz's in the *Prüfung*.

7. Kant, CPR, A146 = B185ff.

8. Kant, CJ, 90; Ak. V, 253ff. The corrections suggested by Benno Erdmann are hardly an improvement on the text.

9. Carl Ludwig Schulze, *Dissertatio Inauguralis exhibens nonulla ad doctrinam judiciis analyticis atque syntheticis spetantia*, 18. [tr.] I have translated *repraesentatio* as *idea*, because in classical Latin it has the same meaning as *Vorstellung*. See my "A Note on Translation" in the Translator's Preface.

10. C. T. Michaelis, "Über Kants Zahlbegriff," *Wissenschaftliche Beilage zum Programm der Charlottenschule*, 8.

11. Ibid., the article as a whole.

12. [tr.] Martin does not perhaps state very clearly what the three questions are, namely: (1) Are mathematical propositions eternal or not? (2) Can all mathematics be reduced to arithmetic? (3) Can all mathematics be reduced to logic? He answers the first in this section and the following one; the second in section A4, and the third in section A5.

13. Bernard Bolzano, *Wissenschaftslehre: Versuch einer neuen Darstellung der Logik*, vol. I, sec. 19.

14. Leibniz, *Philosophische Schriften*, VII, 304.

15. Kant, CPR, A102.

16. Ibid., A103.

17. [tr.] Frau Gottfried Martin has told me that her husband was impressed that such counting marks appeared in prehistoric cave paintings and thought they indicated that the cave painters had a conception of numbers. Cf. Alexander

Marshack, *The Roots of Civilization*, for illustrations of prehistoric numerical markings and a discussion of their significance.

18. Leopold Kronecker, "Über den Zahlbegriff," *Philosophisches Aufsätze*, 266.

19. Kant, CPR, A77 = B102.

20. Ibid., A142f. = B182. [tr.] Martin adds the emphasis indicated by italics.

21. Kant, CJ, 90; Ak. V, 253ff. [tr.] The number system may be in base 10 or base 4; the series of numbers from which all are constructed may be 0, 1, 2, 3, 4, 5, 6, 7, 8, 9 or 0, 1, 2, 3 (or any other series). According to the first system, 10 would be the tenth number, etc., while according to the second, 10 would be the fourth number, etc. (0, 1, 2, 3, 10, . . .). In Martin's example of truck-driver-numbers, the number system is pentadic (0, 1, 2, 3, 4, 10, . . .). The way of thinking of the numbers depends on the base, i.e., on the fundamental way of grouping the units. The passage Martin refers to says:

> Die Einbildungskraft schreitet in der Zusammensetzung, die zur Grössen-vorstellung erforderlich ist, von selbst, ohne dass ihr etwas hinderlich wäre, ins Unendliche fort; der Verstand aber leitet sie durch Zahlbegriffe. [The imagination advances of itself in the assembling which is required for the idea of quantity, on into the infinite without any hindrance; the intellect, however, leads it by numerical concepts.]

The second edition of 1793 actually says *Zusammenfassung* instead of *Zusammensetzung*. Benno Erdmann changed this to be consistent with the usage in the rest of the passage, and the Akademy edition follows him in this. The Vorländer edition does not (see the note there). Martin follows the Akademy version: "Auch die Urteilskraft unterscheidet das Zusammensetzen der Einbildungskraft von der Zusammenfassung im Begriffe."

22. Johann Schultz, *Anfangsgründe*, 55, note 1, "Zur Definition der ganzen Zahl."

23. Isaac Newton, *Arithmetica universalis*, 12, 20.

24. Leibniz, Eduard Bodemann catalogue of manuscripts, (*Bodemann Leibniz-Handschriften der königlichen öffentlichen Bibliothek zur Hannover*), Mss. Phil. VII, B III 24; Louis Courturat, *La Logique de Leibniz d'apres des documents inédits*, 88; Couturat, *Opuscules et Fragment inédits de Leibniz*, 284.

25. Christian Wolff, *Anfangsgründe aller Mathematischen Wissenschaften*, sec. 38.

26. [tr.] See Kant's Prize Essay of 1764, *Inquiry into the Clarity of the Principles of Natural Theology and Morals*, First Reflection, sec. 2, Ak. II, 278.

27. Kant, CPR, B145.

28. Ibid., B147.

29. Kronecker, "Über den Zahlbegriff," *Philosophisches Aufsätze*, 265.

30. Abraham Gotthelf Kästner, *Anfangsgründe, der Arithmetik, Geometrie, ebenen und sphärischen Trigonometrie und Perspectiv*, 6. [tr.] The parenthetical (Algebra, Differential, and Integral Calculus) is added by Martin.

31. Schultz, *Anfangsgründe*, 47.

32. Ibid., 2, 10.
33. Kant, CPR, A717 = B745.
34. Ibid., A142 = B182.
35. Paul Natorp, *Die Logische Grundlagen der exacten Wissenschaften*, 1.
36. Kant, CPR, A164 = B205.
37. Kant, letter to Schultz, 25 November 1788, Ak. X, 554ff. See above in Chapter 3.
38. John Locke, *Essay*, bk. IV, chap. 7, sec. 10.
39. Leibniz, *New Essays*, bk. IV, chap. 7, sec. 10.
40. See the reference below,
41. Leibniz, *Philosophische Schriften*, vol. I ("On Spinoza's *Ethics*"), 141.
42. Hermann Hankel, *Theorie der komplexen Zahlensysteme*, 53.
43. Kant, CPR, B16, and *Prolegomena*, 269.
44. See above, p. 48.
45. See above, p. 48–49. [tr.] Several letters cited by Martin from Kant to Schultz recommend changes in Schultz's manuscripts about Kant's philosophy. The letter of November 1788 is specifically about the synthetic nature of arithmetic.
46. Schultz, *Prüfung,* vol. II, vi.
47. [tr.] See the Appendix for a translation of this section (211–36) of the *Prüfung*.
48. See below, pp. 113–14.
49. Johann August Eberhard, "Von dem Einflusse der sinnlichen Anschauungen auf die Wahrheit und Gewissheit" (On the Influence of Sense Perceptions on Truth and Certainty), *Philosophisches Magazin*, 4 (1791), 69.
50. Lazarus Bendavid, "Deduction der mathematischen Prinzipien aus Begriffen" (Deduction of the Principles of Mathematics from Concepts), *Philosophisches Magazin*, 4 (1791), 421–23.
51. Schultz, *Prüfung,* vol. II, 237, 240.
52. Kant, *On A Discovery*, Ak. VIII, 220.
53. [tr.] The citation is Schultz's footnote e. Johann Albert Heinrich Reimarus (1729–1814), *Ueber die gründe der menschlichen erkenntniss und natürlichen religion*, p. 43 (note).
54. Schultz, *Prüfung,* vol. I, 229–32. [tr.] See the Appendix for the full text. The italicized words are those emphasized in Schultz's original text. Martin does not indicate them.
55. Kant, letter to Schultz, 25 November 1788, Ak. X, 555.
56. Schultz, *Prüfung,* vol. II, 262–3.
57. [tr.] I assume that Schütz is using *intuitus* in the classical Latin sense for the Kantian *Anschauung*.
58. [tr.] Plautus, *Trinummus*, in *Comediae*, ed. W. M. Lindsay, vol. II, Act IV, Sc. 2, 43–44 = lines 885–86. "Si ante lucem ire occipias a meo primio nomine/concubium sit noctis priu' quam ad postremum perveneris."
59. Christian Gottfried Schütz, *Programma de syntheticis mathematicorum pronuntiationibus*, 294.
60. Andreas Metz, *Kurze und deutliche Darstellung des Kantischen Systems*, 49.

61. Andreas Mellin, *Enzyklopädisches Wörterbuch der kritischen Philosophie*, vol. I, 199ff.

62. Ibid., vol. v, 433f.

63. Johann Georg Heinrich Feder, *Über Raum und Causalität zur Prüfung der Kantischen Philosophie*, 48.

64. Dietrich Tiedemann, "Über die Natur der Metaphysik, zur Prüfung von Herrn Professor Kants Grundsätzen," in *Hessische Beiträge zur Gelehrsamkeit und Kunst*, vol. 1, 55.

65. Johann Christoph Schwab, *Ausführliche Erläuterung der von der Königlichen Akademie der Wissenschaften zu Berlin für das Jahr 1791 vorgelegten Frage*: "Welches sind die wirklichen Fortschritte, die die Metaphysik seit Leibnitzens und Wolffens Zeiten in Deutschland in Deutschland gemacht hat?" 168ff.

66. Gottlob Ernst Schulze, *Kritik der theoretischen Philosophie*, vol. I, 177. Schulze also wrote under the name of "Aenesidemus."

67. Ibid., vol. II, 175.

68. Schultz, *Prüfung*, vol. I, 231.

69. Johann Gottlieb Fichte, *Über den Begriff der Wissenschaftslehre oder der sogenannte Philosophie, in Ausgewählte Werke*, ed. F. Medicus, vol. I, 194, note.

70. Carl Friedrich Gauss, letter to Heinrich Schumacher, 1 November 1844, *Briefwechsel zwischen C. F. Gauss und H. C. Schumacher*, ed. C. A. F. Peters, vol. IV, 337. [tr.] Friedrich Wilhelm Joseph Schelling (1775–1854) and Christian Nees von Esenbek (1776–1858).

71. [tr.] Martin's text includes this section (from "This question" to "or G. E. Schulze") as part of Gauss's letter of 1 November 1844, but it is not. Gauss's letter continues:

Read in the history of ancient philosophy what the important men of the time, Plato and others (I make an exception of Aristotle), gave for explanations. But even with Kant it is often not much better; his distinction between analytic and synthetic propositions is in my view one which either just boils down to a triviality or is simply false. . . .

It appears that Martin himself wrote the passage in question, and that it was mistakenly set as extract in the 1972 edition of *Arithmetik und Combinatorik bei Kant*. I have restored it as text.

The reference to Hegel is to his *Wissenshaft der Logik* (1812–16), ed. G. Lasson, vol I, 200.

72. Georg Wilhelm Friedrich Hegel, *Wissenschaft der Logik*, vol. I, 202.

73. Jakob Friedrich Fries, *Neue Kritik der Vernunft*, sec. 65.

74. Friedrich Ernst Daniel Schleiermacher, *Dialektik*, Ak. IV, 2, 563.

75. Aloys Riehl, *Der philosophische Kritizismus und seine Bedeutung für die positive Wissenschaft,*, 443ff.

76. Willi Reichardt, "Kants Lehre von den synthetischen Urteilen a priori und ihre Bedeutung für die Mathematik," *Philosophische Studien*, 4 (1888), 608.

77. [tr.] I have not been able to locate Martin's references precisely. Heinrich Rickert discusses mathematics in all three of the following major works:

Allgemeine Grundlegung der Philosophie (Tübingen, 1921); *Der Gegenstand der Erkenntnis: Einführung in die Tranzendental-Philosophie* (Tübingen, 1915); *Kulturwissenschaft und Naturwissenschaft* (3d ed., Tübingen, 1915).

78. Gottlob Frege, *Die Grundlagen der Arithmetik: Eine logische-mathematische Untersuchung über den Begriff der Zahl*, 7ff.

79. Leonhard Nelson, *Kant und die Nicht-Euklidische Geometrie*, 153.

80. Frege, *Die Grundlagen der Arithmetik*, 6f.

81. Frege, *Die Grundlagen der Arithmetik*, 6.

82. [tr.] Martin adds: I have already examined Hermann Hankel in several places.

83. Paul Mansion, "Gauss contre Kant sur la géométrie non euclidienne," in *Bericht über den 3. internationalen Kongress für Philosophie*, 445.

84. Otto Hölder, *Die mathematische Methode*, 326, sec. 65; sec. 124; sec. 127.

85. Heinrich Behmann, "Sind die mathematischen Urteile analytisch oder synthetisch?" *Erkenntniss*, 4 (May 1934), 6.

86. Ibid., 5.

87. Ibid., 21ff.

88. Ibid., 12.

89. Frantisek Prihonsky, *Neuer Anti-Kant, oder Prüfung der Kritik der reinen Vernunft*, 2.

90. Bolzano, *Wissenschaftslehre*, vol. I, 113.

91. Ibid., vol. I, 216.

92. Ibid., sec. 72.

93. Ibid., sec. 73.

94. Ibid., sec. 73.

95. Ibid., vol. I, sec. 79, 364.

96. Ibid., vol. II, sec. 148, 83.

97. Ibid., vol. I, sec. 65, 288.

98. Ibid., vol. III, sec. 305, 186.

99. Ibid.; and *Beiträge zu einer begründeten Darstellung der Mathematik*, 147.

100. Robert Zimmermann, "Über Kants mathematisches Vorurtheil und dessen Folgen," *Sitzungsbericht der phil.-hist. Classe der Kaiserlichen Akademie der Wissenschaften*, vol. 67, 16.

101. Julius Bergmann, "Über den Satz des zureichenden Grundes," *Zeitschrift für imm. Philosophie, II, Untersuchung über die Hauptmerkmale der Philosophie*, 291ff.

102. Franz Brentano, *Versuch über die Erkenntnis*, from the *Nachlass*, 47.

103. Edmund Husserl, *Formale und transzendentale Logik: Versuch einer Kritik der logischen Vernunft*, 80.

104. Ibid., 86.

105. Ibid., 68.

106. Husserl, *Philosophie der Arithmetik*, 204; *Husserliana: Gesammelte Werke*, vol. XII, 184. [tr.] Husserl's references are to: 1) Friedrich Albert Lange, *Geschichte des Materialismus und Kritik seiner Bedeutung in der Gegenwart* (3d ed., 2 vols., Iserlohn, 1876–77), bk. II (*Geschichte des Materialismus seit Kant*), 119; 2) Zimmermann, "Über Kants mathematisches Vorurtheil und

dessen Folgen," *Sitzungsbericht der phil.-hist. Classe der Akademie der Wissenschaften*, vol. 67, 16ff.

107. Husserl, *Logische Untersuchungen*, vol. II, 1, sec. 12, 254–56.

108. Hermann Ritzel, *Über analytische Urteile: eine Studie zur Phänomenologie des Begriffs*, 333.

109. Hermann Cohen, *Das Prinzip der Infinitesimalmethode und seine Geschichte*.

110. Paul Natorp, *Logische Grundlagen*, 2.

111. Ibid., 130.

112. Ibid., 149.

113. Natorp, "Zu den Logischen Grundlagen der neueren Mathematik," *Archiv für systematische Philosophie*, vol. VII, 177ff., 372ff.

114. Hermann Cohen, *Kants Theorie der Erfahrung*, Preface.

115. Ibid., 514.

116. Ibid., chap. 11, pt. 2.

117. Ibid., 515.

118. Natorp, *Logische Grundlagen*, 45.

119. Ibid., 44ff.

Appendix

1. This account of Johann Schultz's life depends primarily on Otto Liebmann's article, "Johann Schultz," in the *Allgemeine Deutsche Biographie*.

2. *Prüfung der Kantischen Critik der reinen Vernunft,*, pt. 1, *Prüfung der Einleitung*, sec. 4, *Giebt es theoretische Wissenschaften, welche synthetische Urtheile a priori enthalten?*, 211–36.

3. [tr.] (a) The binomial theorem, e.g.: $(a + b)^2 = a^2 + 2ab + b^2$. (b) Sums of series, e.g.: $1 + q + q^2 + q^3 + \ldots = 1/1 - q$ if $|q| \langle 1$. (c) Thomas Harriot (1560–1621). The law of signs only works for real numbers.

4. [tr.] Christian Wolff, *Elementa matheseos universae*.

5. [tr.] The following is Schultz's footnote *d*:

> I do not first have to remember that the algebraist really needs all the same expressions of the infinite and finds them useful in the calculus. Yet one must not draw the conclusion from this that all these expressions, except for subjective reality, which is all that is under discussion with this postulate, also have objective reality as well, i.e., that they are also collectively applicable to real objects. I have gone into detail about what kind of special circumstances it has in my *Versuch einer genauen Theorie des Unendlichen* (Exploration of a Precise Theory of the Infinite, Königsberg and Leipzig, 1788), secs. 32–34.

6. [tr.] The present sec. 4, II, *Die Geometrie besteht aus lauter synthetischen Sätzen a priori* (Geometry Consists of Pure Synthetic Propositions A Priori), 2, *Die geometrischen Postulate und Axiomen sind gleichfalls lauter synthetische Sätze* (The geometrical postulates and axioms are likewise pure synthetic propositions), 65–72.

7. [tr.] The citation is Schultz's footnote *e*: "[Johann Albert Heinrich Reimarus,] *Ueber die gründe d[er] m[enschlichen] Erkenntniss [und natürlichen religion. . . .*], p. 43 (note)."

Bibliography

A. Publications by Gottfried Martin

A FULL BIBLIOGRAPHY of Gottfried Martin's works is to be found in the festschrift for his sixty-fifth birthday by E. Gerresheim, *Einheit und Sein*, ed. I. Heidemann and E. K. Specht, *Kant-Studien*, 57 (1966), 1–3: 400–415. Additions to it are to be found in *Kant-Studien*, 73 (1982): 499–501, compiled by E. Gerresheim and G. Buhl.

Some catalogues mistakenly identify Martin as the author of *Brennende Wunden: Tatsachenbericht über die Notlage der evangelischen Deutschen in Polen*, Berlin-Steglitz, 1931. The author was actually Dr. Richard Kammel, who used the pseudonym "Gottfried Martin" because of the political situation of Germans in Poland at the time.

Articles by Martin cited in this translation are listed in Part C of this bibliography. The following is a list of his books.

Arithmetik und Kombinatorik bei Kant. Dissertation. Freiburg, 1934. Itzehoe, 1938; 2d ed., Berlin, 1972.

Wilhelm von Ockham. Berlin, 1949.

Immanuel Kant, Ontologie und Wissenschaftstheorie. Cologne, 1951; 4th ed., Berlin, 1969. Translated by P. G. Lucas, under the title *Kant's Metaphysics and Theory of Science.* Manchester, 1955, 1961; New York, 1956; Westport, Conn., 1974.

Neuzeit und Gegenwart in der Entwicklung des mathematischen Denkens. Cologne, 1954. Sonderdruck, *Kant-Studien*, vol. 45, 1–4 (1953–54).

Klassische Ontologie der Zahl, Cologne, *Kant-Studien*, Ergänzungsheft, 1956.

Einleitung in die allgemeine Metaphysik. Cologne, 1957, 2d ed., 1958; Stuttgart, 1965. Translated by E. Schaper and I. Leclerc, under the title *An Introduction to General Metaphysics*, London, 1961; New York, 1963.

Leibniz, Logik und Metaphysik. Cologne, 1960; 2d ed., Berlin, 1967. Translated by K. J. Northcott and P. G. Lucas, under the title *Leibniz, Logic and Metaphysics*, Manchester and New York, 1964.

Allgemeine Metaphysik. Berlin, 1965. Translated by D. O'Connor, under the

title *General Metaphysics*. London and New York, 1968.
Sokrates. Reinbeck bei Hamburg, 1967.
Idee und Wirklichkeit der deutschen Universität. Bonn, 1967.
Platon. Reinbeck bei Hamburg, 1969.
Platons Ideenlehre. Berlin, New York, 1973.
Gesammelte Abhandlungen und Vorträge. I, *Kant-Studien*, Ergänzungsheft, 81, 1961.
Edited with D. Löwisch. *Sachindex zu Kants Kritik der reinen Vernunft*. Berlin, 1967.

B. Translations of Descartes, Hegel, Husserl, Kant, and Leibniz Consulted

Descartes, René. *Meditations on First Philosophy*. Translated by Laurence J. Lafleur. Indianapolis, 1960.

―――. *The Philosophical Works*. Translated by Elizabeth S. Haldane and G. R. T. Ross. Cambridge, 1967.

Hegel, G. W. F. *Science of Logic*. Translated by W. H. Johnston and L. G. Struthers. London and New York, 1929.

Husserl, E. *Formal and Transcendental Logic*. Translated by Dorion Cairns. The Hague, 1969.

Kant, I. *Critique of Pure Reason*. Translated by F. Max Müller. London and New York, 1896.

―――. *Critique of Pure Reason*. Translated by Norman Kemp Smith. New York, 1965.

―――. *Prolegomena to Any Future Metaphysics*. Translated by L. W. Beck. New York, 1950.

―――. *Nova dilucidatio* (Appendix to *Kant's Conception of God*). Translated by F. E. England. London, 1929.

―――. *Philosophical Correspondence, 1759–99*. Translated and edited by A. Zweig. Chicago, London, and Toronto, 1967.

―――. *Selected Pre-Critical Writings and Selected Correspondence with Beck*. Translated and edited by G. B. Kerferd and D. E. Walford. Manchester, 1968.

―――. *Kant* (selections). Translated and edited by Gabriele Rabel. Oxford, 1963.

Leibniz, G. W. *Logical Papers*. Translated and edited by G. H. R. Parkinson. Oxford, 1966.

―――. *New Essays Concerning Human Understanding*. Translated by A. C. Langley. New York, 1896.

―――. *New Essays On Human Understanding*. Translated by Peter Remnant and Jonathan Bennett. Cambridge, 1981.

―――. *Philosophical Papers and Letters*. Translated and edited by L. E. Loemker. Chicago, 1965; 2d ed., Dordrecht, 1970. 1965, 2d ed. Dordrecht, 1970

C. Works Cited

The following list comprises the original *Verzeichnis* of Gottfried Martin together with other works consulted by the translator in preparing this English-language edition.

Adickes, Erich. *Kant als Naturforscher*. 2 vols. Berlin, 1924–1925.

Aesop. *Aesopica*. Translated and edited by Ben Edwin Perry. Urbana, Ill., 1952.

———. *Fables*. Translated by V. S. Vernon. New York, 1912.

Aleksandrov, A. D., A. N. Kolmogorov, M. A. Lavrent'ev, eds. *Mathematics: Its Content, Methods, and Meaning*. Translated by S. H. Gould and T. Bartha. Moscow, 1956; Cambridge, Mass., 1963.

d'Alembert, Jean le Rond. *Reflexions sur la cause generale des vents*. Berlin, 1747.

Allison, Henry E. *The Kant-Eberhard Controversy*. An English translation together with supplementary materials and a historical-analytic introduction of Immanuel Kant's *On a Discovery According to which Any New Critique of Pure Reason Has Been Made Superfluous by an Earlier One*. (*Über eine Entdeckung nach der alle neue Kritik der reinen Vernunft durch eine ältere entbehrlich gemacht werden soll*.) Baltimore and London, 1973.

Apollonius of Perga. *Apollonii Pergaei quae Graece exstant cum commentariis Antiquis*. 2 vols. Editit et Latine Interpretatus est I. L. Heiberg. Lipsiae, 1893.

Arnoldt, Emil. *Kritische Exkurse im Gebiete der Kant-Forschung, Gesammelte Schriften*. 5 vols. Edited by O. Schöndörffer, Berlin, 1906–1911.

Bacon, Francis. *Novum Organum*. Edited by Thomas Fowler. Oxford, 1828.

———. *The New Organon*. Translated by James Spedding, Robert Leslie Ellis, and Douglas Devon Heath. Edited by Fulton H. Anderson. Indianapolis, 1960.

Baltzer, Richard. *Die Elemente der Mathematik*. 2 vols. 7th ed. Leipzig, 1885.

Barker, Stephen F. *Philosophy of Mathematics*. Englewood Cliffs, N. J., 1964.

Barrow, Isaac. *Lectiones geometricae*. London. 1670.

———. *The Mathematical Works*. Edited by W. Whewell, London, 1861.

———. See Euclid.

Baudouin de Guémaduc, Armand Henri. *Abhandlung von der Entdeckung eines Trabanten der Venus*, from the French *Remarques sur une quatrième observation du satellite de Venus*. Berlin, 1761.

Beck, Hans. *Einführung in die Axiomatik der Algebra*. Berlin, 1926.

Beck, Jacob Sigismund. *De theoremate Tayloriano, sive delege generali*. Dissertation. Halle, 1791.

———. *Erläuternder Auszug aus den kritischen Schriften des Herrn Professor Kant*. 3 vols, vol. 3 with separate title *Einzig möglicher Standpunkt, aus welchem die kritische Philosophie beurteilt werden muss*. Riga, 1793–1796.

Becker, Oskar. "Phänomenologische Begründung der Geometrie." *Jahrbuch für Philosophie und phänomenologische Forschung*. Edited by E. Husserl, vol. 6, 385–560. Halle, 1923.

———. "Mathematische Existenz." *Jahrbuch für Philosophie und phänomenologische Forschung*. Edited by E. Husserl, vol. 7, pp. 439–809. Halle, 1927.

Beckmann, Petr. *A History of π (Pi)*. Boulder, Colo., 1970; 2d ed., 1971; 3d ed., 1974.

Behmann, Heinrich. "Sind die mathematischen Urteile analytisch oder synthetisch?" *Erkenntnis* 4 (May 1934).

Belidor, Bernard Forest de. *Architectura hydraulica*. 2 parts. Augsburg. Pt. 1, 1740; pt. 2, 1766.

Bell, Eric Temple. *The Development of Mathematics*. New York and London, 1945.

Benacerraf, Paul and Hilary Putnam, eds. *Philosophy of Mathematics*. Oxford, 1964; Englewood Cliffs, N. J., 1964.

Bendavid, Lazarus. "Deduction der mathematischen Prinzipien aus Begriffen," *Philosophisches Magazin*. Edited by J. A. Eberhard. 4 (1791–92): 271–301, 406–23.

Benvenuti, Carol. *Dissertatio physica de lumine*. Vindobonae, 1761.

Bergmann, Julius. "Über den Satz des zureichenden Grundes." *Zeitschrift für imm. Philosophie*, II. *Untersuchungen uber die Hauptmerkmale der Philosophie*. Marburg, 1900.

Bernoulli, Daniel. *Hydrodynamica sive de viribus et motibus fluidorum commentarii*. Argentorati, 1738.

Bernoulli, Jacob. *Ars conjectandi, Opus postumum*. Berlin, 1713.

Bernoulli, Jacques (Jacob or James). *The Doctrine of Permutations and Combinations*. Translated by Frances Masères(?). London, 1795.

Beth, Evert. *The Foundations of Mathematics: A Study in the Philosophy of Science*. Amsterdam, 1965.

———. "Nieuwentyt's Significance for the Philosophy of Science," *Synthese* 9 (1953–5): 447–453.

Bodemann, Eduard. *Bodemann Leibniz-Handschriften der königlichen öffentlichen Bibliothek zur Hannover*. Hannover, 1895.

Boethius. *On Porphyry's Isagoge, Second Commentary, Corpus Scriptorum Ecclesiasticorum Latinorum*. Edited by S. Brandt. Vienna, 1906.

———. "The Second Edition of the Commentaries on the Isagogue of Porphyry." Translated by Richard McKeon. In *Selections from Medieval Philosophers*, I. New York, 1929.

Bolyai, Wolfgang (Farkas). See Gauss.

Bolzano, Bernard. *Beiträge zu einer begründeten Darstellung der Mathematik*. Prague, 1810.

———. *Wissenschaftslehre: Versuch einer neuen Darstellung der Logik*, 4 vols. Sulzbach, 1837; reprint edited by W. Schultz, Leipzig, 1929–31.

———. *Theory of Science*. Translated and edited by R. George. Oxford, 1972.

Borelli, J. A. See Euclid.

von Braunmühl, A. "Trigonometrie, Polygonometrie und Tafeln." In *Vorlesungen über Geschichte der Mathematik*. 2d ed. Edited by Moritz Cantor. Leipzig, 1908; Stuttgart, 1965. Vol. 4, Pt. 23.

Brentano, Franz. *Versuch über die Erkenntnis,* from the Nachlass. Edited by A. Katsil, Leipzig, 1925; 2d ed., edited by F. Mayer-Hillebrand, Hamburg, 1970.

Bretschneider, Carl Anton. *System der Arithmetik und Analysis*. Jena, 1856.

Brouwer, Luitzen Egbertus Jan. "Intuitionism en Formalisme." Inaugural Address at the University of Amsterdam, 1912. In *Wiskunde, Waarheid, Werkelijkheid.* Groningen, 1919. Translated by Arnold Dresden, under the title "Intuitionism and Formalism." *Bulletin of the American Mathematical Society.* 20 (1913): 81–96. In P. Benacerraf and H. Putnam, eds. *Philosophy of Mathematics: Selected Readings.* Oxford, 1964; Englewood Cliffs, N.J., 1964.

Büttner, Christoph Andreas. *Erläuterung der Rechenkunst, Geometrie und Trigonometrie, welche sich in . . . Wolffs Auszügen aus den Anfangsgründen aller mathematischen Wissenschaften befinden.* Stettin and Leipzig, 1754.

Cajori, Florian. *History of Mathematical Notations.* Vol. II. La Salle, Ill., 1929.

Cantor, Moritz. *Vorlesungen über Geschichte der Mathematik.* 3 vols. Leipzig, 1880–98; 2d ed., 4 vols., Leipzig, 1894–1908.

Carnap, Rudolf. "The Logicist Foundations of Mathematics." Translated by Erna Putnam and Gerald J. Massey. In *Philosophy of Mathematics*, edited by P. Benacerraf and H. Putnam. Oxford, 1964; Englewood Cliffs, N.J., 1964. Originally in *Erkenntnis* 2 (1931).

Cassirer, Ernst. *Das Erkenntnisproblem in der Philosophie und Wissenschaft der neueren Zeit.* 3 vols. Berlin, 1906–20; 3d ed., Berlin, 1920–23.

———. *Substanzbegriff und Funktionsbegriff: Untersuchungen über die Grundfragen der Erkenntniskritik.* Berlin, 1910; Darmstadt, 1969.

Cicero. *De Inventione, De Optima Genere Oratorum, Topica.* Latin-English. Translated by H. M. Hubble. Cambridge, Mass., 1949.

Clavius, Christopher. See Euclid.

Cohen, Hermann. *Das Prinzip der Infinitesimalmethode und seine Geschichte.* Berlin, 1883.

———. *Kants Theorie der Erfahrung.* Berlin, 1871; 4th ed., Berlin, 1924.

———. *System der Philosophie.* 3 pts. Berlin, 1889–1904; 2d, 4th eds., Berlin, 1921–23.

Couturat, Louis. *La logique de Leibniz d'après des documents inédits.* Paris, 1901; Hildesheim, 1961.

———. "La philosophie des mathématiques de Kant." *Revue de Métaphysique et de Morale* 12 (1904). Also as appendix to *Les Principes des Mathématiques, avec un appendix sur la philosophie des mathématiques de Kant.* Paris, 1905.

———. *Opuscules et Fragment inédits de Leibniz.* Paris, 1903; Hildesheim, 1966.

Crelle, August Leopold, ed. *Journal für reine und angewandte Mathematik.* Berlin, 1826–55.

Crusius, Christian August. *Entwurf der notwendigen Vernunftwahrheiten, wiefern sie den zufälligen entgegengesetzt werden.* Leipzig, 1745.

———. *Weg zur Gewissheit und Zuverlässigkeit der menschlichen Erkenntnis.* Leipzig, 1747.

Delone, B. N. "Algebra: Theory of Algebraic Equations." In *Mathematics: Its Content, Methods, and Meaning*, edited by A. D. Aleksandrov, A. N.

Kolmogorov, and M. A. Lavrent'ev; translated by S. H. Gould and T. Bartha. Moscow, 1956; Cambridge, Mass., 1963.

Des Cartes, Renatus. *Geometria*, anno 1637 Gallice edita. Edited by Schooten. Lugduni Batavorum, 1649.

Descartes, René. *Meditations on First Philosophy.* Translated by Laurence J. Lafleur. Indianapolis, 1960.

———. *Discourse on Method, Optics, Geometry, and Meteorology.* Translated by Paul J. Olscamp. New York, 1964.

———. *The Geometry.* Translated by David Eugene Smith and Marcia L. Latham. New York, 1954.

———. *Meditationes de Prima Philosophia.* In *Oeuvres*, vol. 7, edited by Charles Adam and Paul Tannery. French version, vol. 9. Paris, 1964.

———. *La Geometrie.* In *Oeuvres*, vol. 6, edited by Charles Adam and Paul Tannery. Paris, 1965.

———. *The Philosophical Works.* 2 vols. Translated by Elizabeth S. Haldane and G. R. T. Ross. Cambridge, 1967.

Dickenson, Edmund. *Physica vetus et vera sive tractatus de naturali veritate hexaemeri mosaici.* Roterdami, 1703.

Dieterich, Konrad. *Kant und Newton.* Tübingen, 1877.

Dilthey, Wilhelm. "Neue Kanthandschriften." *Kant-Studien* 3 (1899).

Doppelmayr, Johann Gabriel. *Eröffnung der neuen Mathematischen Werck-Schule . . . , in welcher sowohl die Zubereitung als der Gebrauch verschiedener anderer Mathematischen . . . Instrumenten . . . erkläret werden.* Nürnberg, 1717.

Eberhard, Johann August. "Von dem Einflusse der sinnlichen Anschauungen auf die Wahrheit und Gewissheit." *Philosophisches Magazin 4* (1791–92): 68–83. Brussels, 1968.

———. "Ueber die apodiktische Gewisheit." *Philosophisches Magazin 2* (1789–90): 129–85. Brussels, 1968.

———, ed. *Philosophisches Magazin.* 4 vols. Halle, 1788–92. Brussels, 1968.

Eberhard, Johann Peter. *Beiträge zur Mathesi Applicata.* Halle, 1757.

———. *Erste Gründe der Naturlehre.* Leipzig, 1753; 4th ed., Halle, 1774.

Edwards, Paul, ed. *The Encyclopedia of Philosophy.* 8 vols. New York and London, 1967.

Engel, Friedrich. "Grassmans Leben." In H. Grassmann, *Gesammelte mathematische und physikalische Werke.* Vol. 3. 1911.

Enriques, Federigo. *Zur Geschichte der Logik.* Leipzig, 1927.

Erdmann, Benno. *Beiträge zur Geschichte und Revision des Textes von Kants Kritik der reinen Vernunft: Anhang zur fünften Auflage der Ausgabe* (Appendix to the fifth edition). Berlin, 1900.

Erxleben, Johann Christian Polycarp. *Anfangsgründe der Naturgeschichte, zum Gebrauch Akademischer Vorlesungen.* 2 pts. Göttingen and Gotha, 1768.

Euclid. *Euclidis elementorum libri xv.* Edited by C. Clavius. Coloniae, 1591.

———. *Euclides restitutus.* Edited by J. A. Borelli. Pisa, 1658.

———. *Euclidis elementorum libri xv.* Edited by Isaac Barrow. London, 1659.

———. *Euclidis opera omnia.* Edited by J. L. Heiberg and H. Menge. 1883.

————. *Die Elemente, Gesamtausgabe.* 7 vols. Edited by J. L. Heiberg and H. Menge; translated by Clemens Thaer. Leipzig, 1883–95; Darmstadt, 1971.

————. *The Thirteen Books of Euclid's Elements.* Translated and edited by T. L. Heath. 1908. 2d ed., with additions, 1925; 3 vols., New York, 1956.

————. *Elementa.* Edited by E. S. Stamatis. Leipzig, 1969.

Euler, Leonhard. *Mechanica sive motus scientia analytice exposita.* 2 vols. Petersburg, 1736; German ed. J. Ph. Wolfers, 3 vols. in 4 pts., Greisswald, 1848–53.

————. *Institutiones calculi differentialis.* 2 vols. Petersburg, 1755, 1804; German ed., ed. J. A. C. Michelsen, 3 vols. and 1 supplementary vol., Berlin and Libau, 1790–98.

————. *Institutiones calculi integralis.* 3 vols. Petersburg, 1768–70; 3d ed., 4 vols., 1824–47; German ed. Salomon, Vienna, 1828–30.

————. *Anleitung zur Algebra.* 2 vols. Petersburg, 1770; *Auszug daraus* by Ebert, 2 pts., Berlin, 1801.

————. *Vollständige Anleitung zur Differential-Rechnung.* Translated and edited by J. A. C. Michelsen from the Latin with notes and appendices, pt. I. Berlin and Libau, 1790.

————. *Opera omnia.* Edited by the Schweizerische Naturforschende Gesellschaft. Leipzig and Berlin, 1911.

Feder, Johann Georg Heinrich. *Über Raum und Causalität zur Prüfung der Kantischen Philosophie.* Göttingen, 1787.

Fichte, Johann Gottlieb. *Über den Begriff der Wissenschaftslehre oder der sogenanten Philosophie.* Weimar, 1794.

————. *Ausgewählte Werke.* 6 vols. Edited by F. Medicus. Vol. 1, 2d ed. Darmstadt, 1962.

Fowler, H. W. *A Dictionary of Modern English Usage.* 2d ed., Oxford, 1983.

Frege, Gottlob. *Die Grundlagen der Arithmetik: Eine logische-mathematische Untersuchung über den Begriff der Zahl.* Breslau, 1884; 2d ed., Oxford, 1953; Darmstadt, 1970. Translated by J. L. Austin, under the title *The Foundations of Arithmetic.* With German text. Oxford, 1950; 2d ed., Oxford, 1953; English only ed., Oxford, 1980.

(Friedrich II), King of Prussia. *Des Königs von Preussen Majestät Unterricht von der Kriegskunst an seine Generals.* Frankfurt and Leipzig, 1761.

Fries, Jakob Friedrich. *Neue oder anthropologische Kritik der Vernunft.* 3 vols. Heidelberg, 1807; 2d ed., 1828–31.

————. *Die mathematische Naturphilosophie nach philosophischen Methode bearbeitet.* Heidelberg, 1822.

Funke, Gerhard, and Joachim Kopper. "Gottfried Martin." *Kant-Studien* 63 (1972).

Furtenbach, Joseph der Ältere. *Mannhafter Kunst-Spiegel, oder Continuatio, und fortsetzung allerhand Mathematisch- und Mechanisch-hochnutzlich-so wol auch sehr erfrölichen delectationen, und respective im Werck selbsten experimentirten freyen Künsten.* Augsburg, 1663.

Galilei, Galileo. *Systema cosmicum.* Lugduni Batavorum, 1699.

Gardner, Martin. *Logic Machines, Diagrams, and Boolean Algebra*. New York, 1968.

Gassendi, Pierre. *Institutio astronomica iuxta hypotheseis tam veterum, quam Copernici et Tychonis*. Paris, 1647.

Gauss, Carl Friedrich. *Briefwechsel zwischen C. F. Gauss und H. C. Schumacher*. Edited by C. A. F. Peters. 6 vols. Altona, 1860–65. In *Werke*, Ergänzugsreihe, vol. 5, parts I–III. Hildesheim and New York, 1975.

———. *Werke*. 12 vols. Edited by Ernst Schering et al. for the Göttingen Akademie der Wissenschaften. Göttingen, 1863–1929. Hildesheim and New York, 1973.

———. *Briefwechsel zwischen Carl Friedrich Gauss und Wolfgang Bolyai*. Edited by Franz Schmidt and Paul Stäckel. Leipzig, 1899.

———. *Nachträge zum Briefwechsel zwischen Carl Friedrich Gauss und Heinrich Christian Schumacher*. Edited by Theo Gerardy. Göttingen, 1963.

George, Rolf. "*Vorstellung* and *Erkenntnis* in Kant." In *Interpreting Kant*, ed. Moltke S. Gram. Iowa City, 1982.

Gerresheim, Eduard. Bibliography der Veröffentlichungen, Festschrift for Gottfried Martin on his 65th birthday, "Einheit und Sein," ed. I. Heidemann and E. K. Specht, *Kant-Studien*, 57 (1966): 400–415.

Gillispie, Charles Coulston, ed. *Dictionary of Scientific Biography*. 16 vols. New York, 1970–80.

Gow, James. *A Short History of Greek Mathematics*. New York, 1884, 1923.

Gram, Moltke S., ed. *Interpreting Kant*. Iowa City, 1982.

———. "The Sense of a Kantian Intuition." In *Interpreting Kant*, ed. Moltke S. Gram. Iowa City, 1982.

Grassmann, Hermann Günter. *Die Wissenschaft der extensiven Grössen oder die Ausdehnungslehree*. Leipzig, 1844; 2d ed., 1878.

———. *Lehrbuch der Arithmetik und Trigonometrie für Höhere Lehranstalten*. 2 pts. Berlin, 1861, 1865.

———. *Gesammelte mathematische und physikalische Werke*. Edited by F. Engel for the Königliche Sächsische Gesellschaft der Wissenschaften. Leipzig, 1894.

Gravesande, Guilelm Jakob. *Matheseos universalis elementa*. Lugduni Batavorum, 1727.

Günther, S. (A. W. Siegmund). "Geschichte der Mathematik." In *Vorlesungen über Geschichte der Mathematik*, ed. M. Cantor. Vol. IV (1759–99). Leipzig, 1908; Stuttgart, 1965.

Hadaly von Hada (Hadali de Hada), Carl (Karl). *Anfangsgründe der Mathematik*. Pressburg, 1789.

Hales, Stephen. *Statick der Gewächse . . . mit einer Vorrede . . . von Wolff* (translated from the English *Vegetable Staticks*). Halle, 1748.

Hamilton, William Rowan. "Theory of Conjugate Functions, or Algebraic Couples; with a Preliminary and Elementary Essay on Algebra as the Science of Pure Time." *Transactions of the Royal Irish Academy*, XVII (1837): 293–422.

———. *Lectures on Quaternions*. Dublin, 1853.

—. *Elements of Quaternions*. London, 1886; 2d ed., 2 vols., 1899
New York, 1969; German ed., ed. Glan, 2 vols., 1881–84.

—. *The Mathematical Papers*. *3 vols*. Edited by H. Halberstam and
Ingram for the Royal Irish Academy. Cambridge, 1931–67.

Hankel, Hermann. *Theorie der komplexen Zahlensysteme*. Leipzig, 186'

—. *Zur Geschichte der Mathematik in Alterthum und Mittelalter*. Le
1875.

Hankins, Thomas L. *Sir William Rowan Hamilton*. Baltimore and Lo
1980.

Hanov, Michael Christoph. *Philosophiae naturalis sive physicae dogmati(*
vols. Halle, 1762, 1765.

Hausen, Christian August. *Elementa matheseos*, pars prima. Lipsiae, 17

Heath, Thomas L. *A History of Greek Mathematics*. 2 vols. Oxford, 192

—. See Euclid.

Hegel, Georg Wilhelm Friedrich. *De orbitis planetarum*. Jena, 180
Sämtliche Werke, ed. Hermann Glockner, vol. 1, 2d ed. Stuttgart, 194

—. *Wissenschaft der Logik*. Heidelberg, 1812–16. 2 vols. Edited k
Lasson. 2d ed., Leipzig, 1934; 4th ed., Hamburg, 1967–69.

—. *Science of Logic*. Translated by W. H. Johnston and L. G. Strui
London and New York, 1929, 1951.

Heiberg, Johan Ludvig. *Literaturegeschichtliche Studien über Euklid*. Lei
1882.

—. "Jahresberichte: Griechische und römische Mathematik." *Philol(*
2 (1884).

—, ed. *Apollonii Pergaei quae Graece Exstant cum commentariis A*
uis. Lipsiae, 1893.

—. See Euclid.

Helmholtz, Hermann Ludwig Ferdinand von. *Schriften zur Erkenntnisthe(*
Edited by P. Hertz and M. Schlick. Berlin, 1921. Translated by Malcol
Lowe and edited by Robert S. Cohen and Yehuda Elkana, under the
Epistemological Writings. Boston, 1977.

—. *Selected Writings*. Translated and edited by Russell Kahl. Middlet(
Conn., 1971.

—. "Zählen und Messen, erkenntnistheoretisch betrachtet."
losophische Aufsätze Eduard Zeller za seinem fünfzigjährigen Do(
jubiläum gewidmet, (festschrift for Eduard Zeller), pp. 17–52. Lei|
1887. *Wissenschaftliche Abhandlungen*. Vol. 3, pp. 356–91. Leipzig, 1
It is available in several English translations: *Combining and Measu(*
trans. Charlotte Lowe Bryan, New York, 1930; "Numbering and Measu
from an Epistemological Viewpoint," in *Epistemological Writings*,
Cohen and Elkana; "An Epistemological Analysis of Counting and Meas
ment," in *Selected Writings*, ed. Kahl.

Hentsch, Johann Jakob. *Philosophia mathematica complectens metho(*
cogitandi, nec non scientiam rerum universalem ex Euclide restitui
Conamina duo priora. Editio secunda auctior et emendatior. Lipsiae, 1'

Hevel, Johann. *Prodromus astronomiae*. Gedani, 1690.

Heyting, Arend. "The Intuitionist Foundations of Mathematics." Translate(

Erna Putnam and Gerald J. Massey. In *Philosophy of Mathematics*, ed. P.
  Benacerraf and H. Putnam. Oxford, 1964; Englewood Cliffs, N. J., 1964.
  Originally in *Erkenntnis*, 2 (1931).
Hilbert, David. *Grundlagen der Geometrie*. Leipzig, 1899; 11th ed., Stuttgart,
  1972.
————. "Axiomatisches Denken." *Mathematisches Annalen*. 78 (1908).
Hinske, Norbert, and Wilhelm Weischedel. *Kant-Seitenkonkordanz*. Darm-
  stadt, 1970
Hintikka, Jaakko. "Are Logical Truths Analytic?" *Philosophical Review* 74
  (1965): 178–203.
————. "Kant and The Tradition of Analysis." In *Deskription, Existenz, und
  Analytizität*, ed. P. Weingartner. Munich, 1966. Also as chap IX in *Logic,
  Language-Games, and Information*. Oxford, 1973.
————. "Kant on Mathematical Method." In *Kant Studies Today*, ed. L. W.
  Beck. La Salle, Ill., 1969.
————. "On Kant's Notion of Intuition (Anschauung)." In *The First Critique:
  Reflections on Kant's Critique of Pure Reason*, ed. Terence Penelhum and J. J.
  MacIntosh. Belmont, Calif., 1969.
Hintikka, Jaakko, and Unto Remes. *The Method of Analysis: Its Geometrical
  Origin and Its General Significance*. Boston, 1974.
Hölder, Otto. *Die Arithmetik in strenger Begründung, Programmabhandlung
  der Philosophischen Fakultät zu Leipzig*. 1914; 2d ed., Berlin, 1929.
————. *Die mathematische Methode*. Berlin, 1924.
Holland, Georg Jonathan. *Abhandlung über die Mathematik*. Tübingen, 1764.
Husserl, Edmund. *Philosophie der Arithmetik*. Halle, 1891. *Husserliana*, vol.
  12, The Hague, 1970.
————. *Formale und transzendentale Logik: Versuch einer Kritik der logis-
  chen Vernunft*. Halle, 1929. Translated by Dorion Cairns, under the title
  Formal and Transcendental Logic. The Hague, 1969.
————. *Logische Untersuchungen*. 3d ed., vol. II, pts. 1 and 2. Halle, 1922.
  Translated under the title *Logical Investigations*. London, 1970.
Hutton, Charles. *Mathematical and Philosophical Dictionary*. London, 1795; 2
  vols., Hildesheim and New York, 1973.

Irmscher, Hans Dietrich, ed. *Immanuel Kant, Aus den Vorlesungen der Jahre
  1762 bis 1764. Aus Grund der Nachschriften Johann Gottfried Herders.
  Kant-Studien*. Ergänzungsheft, 88. Cologne, 1964.

Jakob, Ludwig Heinrich. *Prüfung der Mendelssohnschen Morgenstunden oder
  aller spekulativen Beweise für das Daseyn Gottes in Vorlesungen . . . Nebst
  einer Abhandlung von Herrn Professor Kant*. Leipzig, 1786.
————. *Grundriss der allgemeinen Logik und kritischen Anfangsgründe zu
  einer allgemeinen Metaphysik*. Halle, 1788; 4th ed., 1810.
Jesper, Johann. *Rechen-Buch . . . In welchen der Algorithmus in gantzen und
  gebrochenen Zahlen . . . zur Recreation deutlich erkläret*. Königsberg,
  1682.
Jones, William. *Synopsis palmariorum matheseos*. London, 1706.

Kant, Immanuel. *Gesammelte Schriften*. Edited by the Königlich Preussische Akademie der Wissenschaften and by the Deutsche Akademie der Wissenschaften in Berlin. 23 vols., Berlin and Leipzig, 1900–55.

————. *Kants handschriftlicher Nachlass, Kants gesammelte Schriften*. Edited by the Königlich Preussische Akademie der Wissenschaften and by the Deutsche Akademie der Wissenschaften in Berlin, Section 3 in 11 parts, ed. E. Adickes, F. Berger, A. Buchenau, and G. Lehmann. Berlin and Leipzig, 1923–1955.

————. *Sämtliche Werke*. 9 vols. and 1 Supplementary vol. in 10 pts. Edited by Karl Vorländer et al., Leipzig und Hamburg, 1904.

————. *Aus den Vorlesungen der Jahre 1762 bis 1764. Auf Grund der Nachschriften Johann Gottfried Herders*. Edited by H. D. Irmscher. *Kant-Studien*, Ergänzungsheft 88. Cologne, 1964.

————. *Critique of Pure Reason*. 1781; 2d ed. 787. Translated by F. Max Müller. London and New York, 1896.

————. *Critique of Pure Reason*. (1st and 2d eds.). Translated by Norman Kemp Smith. London, 1929; New York, 1965.

————. *Fortschritte der Metaphysik*. In *Kleinere Schriften zur Logik und Metaphysik*. Edited by Karl Vorländer. Pt. 3. In *Philosophische Bibliothek* series, 46c. Leipzig, 1921.

————. *Kant*. Selections. Translated and edited by Gabriele Rabel. Oxford, 1963.

————. *Kritik der reinen Vernunft*. 1st and 2d eds. Edited by Albert Görland. Berlin, 1923; edited by Raymund Schmidt, 2d ed. Leipzig, 1930.

————. *Logic*. Edited by Robert Hartman and Wolfgang Schwarz. Indianapolis, 1974.

————. *A New Exposition of the First Principles of Metaphysical Knowledge*, from the Latin *Principiorum primorum cognitionis metaphysicae Nova Dilucidatio*, 1755. Translated by F. E. England, as Appendix to his *Kant's Conception of God*. London, 1929.

————. *On a Discovery According to which Any New Critique of Pure Reason Has Been Made Superfluous by an Earlier One*, from the German *Über eine Entdeckung nach der alle neue Kritik der reinen Vernunft durch eine ältere entbehrlich gemacht werden soll*. 1790. Translated by Henry E. Allison, in *The Kant-Eberhard Controversy*. Baltimore and London, 1973.

————. *Philosophical Correspondence*, 1759–99. Translated and edited by A. Zweig. Chicago, London, Toronto, 1967.

————. *Prolegomena zu einer jeden künftigen Metaphysik die als Wissenschaft wird auftreten*. 1783. Edited by Karl Vorländer. 6th ed., Leipzig, 1920.

————. *Prolegomena to Any Future Metaphysics*. Translated by L. W. Beck. New York, 1950.

————. *Selected Pre-Critical Writings and Correspondence*. Translated and edited by G. B. Kerferd and D. E. Walford, Manchester, 1968.

Karsten, Wenceslaus Joann Gustav. *Mathesis theoretica elementaris atque sublimior*. Rostock and Greifswald, 1760.

————. *Mathematische Abhandlungen*. Halle, 1786.

Kästner, Abraham Gotthelf. *Anfangsgründe der Arithmetik, Geometrie, ebenen und sphärischen Trigonometrie und Perspectiv.* Göttingen, 1758; 5th ed., Göttingen, 1792; 6th ed., Göttingen, 1800.

―――. *Anfangsgründe der angewandten Mathematik.* 3 pts. Göttingen, 1759–61; 4th ed., Göttingen, 1792.

Kiesewetter, Johann Gottfried Karl Christian. *Die ersten Anfangsgründe der reinen Mathematik.* Berlin, 1799; 4th ed. 1818.

Kline, Morris. *Mathematical Thought from Ancient to Modern Times.* New York, 1972.

―――. *Mathematics: The Loss of Certainty.* Oxford, New York,Toronto, Melbourne, 1980.

Klügel, Georg Simon. *Mathematisches Wörterbuch.* 3 vols. Leipzig, 1803–1808; 2 vols. and 2 Supplements with additions by Mollweide and Grunert, Leipzig, 1823–36.

Köbel, Jacob. *Geometrey: von künstlichem Feldmessen.* Frankfurt, 1578.

Kolmogorov, A. N. See Aleksandrov, A. D.

Kopper, Joachim. See Gerhard Funke.

Körner, Stephan.*Kant.* Harmondsworth, 1955; New Haven, 1982.

Kratzenstein, Christian Gottleib. *Abhandlung von dem Einfluss des Mondes in die Witterungen und den menschlichen Körper.* Halle, 1747.

Kronecker, Leopold. "Über den Zahlbegriff." *Philosophische Aufsätze.* Dedicated to Eduard Zeller. Leipzig, 1887.

Lagrange, Joseph Louis de. *Théorie des fonctions analytiques, contenant les principes du calcul différentiel.* Paris, 1788.

―――. *Mecanique analytique.* Paris, 1788.

Lambert, Johann Heinrich (Jean Henri). *Philosophische Schriften.* 6 vols. Edited by Hans-Werner Arndt. Hildesheim, 1965.

―――. *Die freye Perspektive.* Zurich, 1759.

―――. *Neues Organon, oder Gedanken über die Erforschung und Beziehung des Wahren und dessen Unterscheidung von Irrtum und Schein.* 2 vols. Leipzig, 1764.

―――. "Mémoire sur quelques propriétés remarquables des quantités transcendentes circulaires et logarithmique." *Histoire de l'Academie Royale des Sciences et des Belles Lettres de Berlin.* Vol. 17. 1768. Translated in part in *A Source Book in Mathematics, 1200–1800,* ed. D. J. Struik. Cambridge, Mass., 1969.

―――. *Anlage zur Architektonik, oder Theorie des Einfachen und Ersten in der philosophischen und mathematischen Erkenntnis.* 2 vols. Riga, 1771.

―――. *Deutscher gelehrter Briefwechsel.* 5 vols. Edited by Johann Bernoulli. Berlin, 1781–85.

Lange, Friedrich Albert. *Geschichte des Materialismus und Kritik seiner Bedeutung in der Gegenwart.* 2 vols. Iserlohn and Leipzig, 1866; with intro. and critical appendix by Hermann Cohen, Leipzig, 1902. Translated by E. C. Thomas, under the title *History of Materialism.* 3 vols. London, 1877–79; 1 vol., with intro. by Bertrand Russell, Edinburgh, 1925.

Lavrent'ev, M. A. See Aleksandrov, A. D.

Legendre, Adrien Marie. *Essai sur la théorie des nombres*. 2 vols. Paris, 1798; German tr. Maser, Leipzig, 1886.

Leibniz, Gottfried Wilhelm. *Mathematische Schriften*. 7 vols. Edited by C. I. Gerhardt. Berlin and Halle, 1849–63; Hildesheim, 1962.

———. *Philosophische Schriften*. 7 vols. Edited by C. I. Gerhardt. Berlin, 1875–90; Hildesheim, 1960–61, 1965.

———. *Sämtliche Schriften und Briefe*. Edited by the Preussische Akademie der Wissenschaften and by the Deutsche Akademie der Wissenschaften in Berlin. Berlin, 1923.

———. *Hauptschriften zur Grundlegung der Philosophie*. 2 vols. Translated by Artur Buchenau and edited by Ernst Cassirer. Leipzig, 1904, 1906; 3d ed., Hamburg, 1966.

———. *Logical Papers*. Translated and edited by G. H. R. Parkinson. Oxford, 1966.

———. *Neue Abhandlung über den menschlichen Vernunft,* from the French *Nouveaux Essais sur L'Entendement Humain*. Translated by Artur Buchenau and edited by Ernst Cassirer. 3d ed., Leipzig, 1915; 4th ed., Hamburg, 1971.

———. *New Essays Concerning Human Understanding*, from the French *Nouveaux Essais sur l'Entendement Humain*. Translated by A. G. Langley. La Salle, Ill., 1949.

———. *New Essays on Human Understanding*. Translated and edited by Peter Remnant and Jonthan Bennett. Cambridge, 1981.

———. *Nouveaux Essais sur L'Entendement Humain*. In *Philosophische Schriften*, ed. C. I. Gerhardt. Vol. V; Ak. ed., Sechste Reihe, Vol. VI.

———. *Opuscules et fragments inédits*. Edited by L. Couturat. Paris, 1903; Hildesheim, 1961, 1966.

———. *Philosophical Papers and Letters*. Translated and edited by L. E. Loemker. Chicago, 1956; 2d ed., Dordrecht, 1969.

Liebmann, Otto. "Johann Schultz." *Allgemeine Deutsche Biography*. 1891; Berlin, 1971.

Lilienthal, Johann Samuel. *Beschreibung einer leichten und geschwinden Methode, den genauen Inhalt aller krummer und geradelinigen Figuren zu erforschen*. Königsberg, 1759.

Lindemann, Ferdinand (or C. L. F.). "Ueber die Zahl Pi." *Mathematische Annalen*. Vol. XX. 1882.

Lipps, Gottlob Friedrich. *Untersuchung über die Grundlagen der Mathematik, Philosophische Studien*. Vol. 9–12. Edited by W. Wundt.

Locke, John. *An Essay Concerning Human Understanding*. 1690; ed. A. C. Fraser, Oxford, 1894; New York, 1959. Translated by Carl Winckler, under the title *Versuch über den menschlichen Verstand*. 2 vols. Leipzig, 1911–13.

Lull, Ramón (Lullus, Raymondus). *Ars Generalis Ultima*. 1308; Palma Malorca, 1645; Frankfort on Main, 1970.

———. *Ars Brevis*. Palma Malorca, 1669; Frankfort on Main, 1970.

Maclaurin, Colin. *A Treatise of Algebra*. London, 1748.

Magnus, Ludwig Immanuel. *Sammlung von Aufgaben und Lehrsatzen aus der Analytischen Geometrie*. Berlin, 1833.

Mairan, Jean Jacques Dourtous de. *Abhandlung vom Eise,* from the French *Dissertation sur la Glace, ou explication physique de la formation de la Glace et de ses divers Phenoménes, etc.* Leipzig, 1752.

Mansion, Paul. "Gauss contre Kant sur la géometrie non euclidienne." *Bericht über den 3. internationalen Kongress für Philosophie,* Heidelberg, 1908. Edited by Th. Elsenhans. Heidelberg, 1909.

Marquardt, Conrad Theophil. *Elementa astrognosiae.* Regiomontani, 1734.

Marshack, Alexander. *The Roots of Civilization.* New York, 1972.

Martin, Gottfried. Festschrift for Gottfried Martin for his 65th birthday. "Einheit und Sein." Edited by I. Heidemann and E. K. Specht. *Kant-Studien* 57 (1966): 1–415.

———. "Herder als Schüler Kants. Aufsätze und Kolleghefte aus Herders Studienzeit." *Kant-Studien* 41 (1936): 294–306.

———. "Die mathematischen Vorlesungen Kants." *Kant-Studien* 58 (1967): 58–62.

Maupertuis, Pierre Louis Moreau de. *Versuch von der Bildung des Körpers,* from the French *Essais sur la formation des corps organisés.* Leipzig, 1761.

Mayr, Alois (Aloys). *Untersuchung über die wissenschaftliche Methode.* Würzburg, 1845.

McKeon, Richard, ed. *Selections from Medieval Philosophers.* 2 vols. New York, 1930, 1958.

Meier, Georg Friedrich. *Vernunftlehre.* 1752; 2d ed., Halle, 1762.

Meissner, Heinrich Adam. *Philosophisches Lexicon aus Christian Wolffs Sämtlichen Deutschen Schriften.* Bayreuth and Hof, 1737; Düsseldorf, 1970.

Mellin, Georg Samuel Albert. *Enzyklopädisches Wörterbuch der kritischen Philosophie.* 6 vols. in 11 pts. Jena and Leipzig, 1797–1804.

———. *Kunstsprache der kritischen Philosophie.* Jena and Leipzig, 1798.

Menge, H. See Euclid.

Mendelssohn, Moses. *Morgenstunden oder über das Dasein Gottes.* Berlin, 1785.

Metz, Andreas. *Kurze und deutliche Darstellung des Kantischen Systems.* Bamberg, 1795.

Meyer, Wilhelm Franz. "Kant und das Wesen des Neuen in der Mathematik." In *Zur Erinnerung an Immanuel Kant.* Essays on the Occasion of the Centennial of Kant's Death. Halle, 1904.

Michaelis, Carl Theodor. "Über Kants Zahlbegriff." *Wissenschaftliche Beilage zum Programm der Charlottenschule.* Berlin, 1884.

Michelsen, Johann Andreas Christian. *Anleitung der Geometrie in Briefen.* Vol. 1. Berlin, 1790.

Mittelstrass, Jürgen. "The Philosopher's Conception of *Mathesis Universalis* from Descartes to Leibniz." *Annals of Science* 36 (1979): 593–610.

———. "Leibniz and Kant on Mathematical and Philosophical Knowledge." Unpublished paper given at the Leibniz Conference, University of Toronto, November 1982.

Möbius, August Ferdinand. *Der baryzentrische Kalkül.* Leipzig, 1827.

Müller, C. H. *Studien zur Geschichte der Mathematik in Göttingen, Abhandlungen zur Geschichte der mathematischen Wissenschaften.* Vol. 18. 1904.

Müller, Johann Ulrich. *Untrüglicher Stunden-Weise; Das ist: Eine deutliche und curiose Beschreibung aller der Zeit üblichen Sonnen-Uhren.* Ulm, 1712.
Murhard, Friedrich Wilhelm August. *System der Elemente der allgemeinen Grössenlehre.* Lemgo, 1798.

Natorp, Paul. *Die logischen Grundlagen der exakten Wissenschaften.* 2d ed., Leipzig, 1921.
———. "Zu den logischen Grundlagen der neueren Mathematik." *Archiv für systematische Philosophie* 7 (1901): 177ff., 372ff.
Nelson, Leonard. *Kant und die Nicht-Euklidische Geometrie. Das Weltall*, VI, 1906. Monograph in series *Vorträge und Abhandlungen*, 13. Berlin, 1906.
Newton, Isaac. *Philosophiae naturalis principia mathematica.* London, 1687; 3d ed., 1726; 3d ed. with variant readings, ed. A. Koyré and I. Cohen, 2 vols., Cambridge, Mass., 1972.
———. *Philosophiae naturalis principia mathematica.* German ed. J. Ph. Wolfers, Berlin, 1872; 2d ed., Darmstadt, 1963.
———. *Mathematical Principles of Natural Philosophy and The System of the World.* Translated by A. Motte. 1729; revised by F. Cajori, Berkeley and Los Angeles, 1966.
———. *Arithmetica universalis.* Cambridge, 1707; Amsterdam, 1761; London, 1845.
———. *Optice . . . Libri tres.* Londini, 1719; English ed., London, 1704; German ed., Leipzig, 1898.
———. *Opticks.* Based on 4th ed., London, 1730; New York, 1952.
Nieuwentyt (Nieuwentijt), Bernard. *Het regt Gebruik der Weltbeschouningen, ter overtuiginge van ongodisten en ongelovigan aangetoont.* Amsterdam, 1716.
———. *Gronden van zekerheid af de regte betoogwyze der wiskundigen so in het denkbeeldige als in het skelijke: ter weeklegging van Spinosa denkbeelidig samentstil; in ter aanleiding van eene sekere sakelyke wysbegeerte, aangetoont door.* Amsterdam, 1720.
Novalis [Friedrich Leopold Freiherr von Hardenberg]. *Schriften und Briefe.* 4 vols. Edited by P. Kluckhohn and R. Samuel. Leipzig, 1929; 2d ed., with additions, Stuttgart, 1960; 3d ed., revised, edited by P. Kluckhohn and R. Samuel and co-edited by H. J. Mähl and G. Schulz, Stuttgart, 1983.

Ohm, Martin. *Elementar-Zahlenlehre.* Erlangen, 1816.
———. *Kritische Beleuchtungen der Mathematik.* Berlin, 1819.
———. *Versuch eines vollkommen consequenten Systems der Mathematik.* 6 pts., 2d ed., Berlin, 1828–32.

Paccioli, Luca (or Pacioli or de Borgo). *Summa de Arithmetica, geometria, proportioni: et proportionalita.* 1494; Toscolano, 1523.
Pappus. *Pappi Alexandrini Collectionis.* Latin-Greek. Edited by Fridericus Hultsch. Berlin 1877; Amsterdam, 1965.
———. *Der Sammlung des Pappus von Alexandrien.* Bks. VII-VIII, German-Greek ed. Edited by C. I. Gerhardt. Halle, 1871.

Parkinson, G. H. R. "Introduction to Leibniz." In *Logical Papers*. Oxford, 1966.

Parsons, Charles. "Kant's Philosophy of Arithmetic." In *Philosophy, Science, and Method: Essays in Honor of Ernest Nagel*, ed. S. Morgenbesser, P. Suppes, and M. White. New York, 1969.

Pascal, Blaise. *Oeuvres Complètes*. Edited by Louis Lafuma. Paris, 1963.

Paton, Herbert James. *Kant's Metaphysic of Experience: A Commentary on the First Half of the Kritik der reinen Vernunft*. New York, 1936.

Peano, Giuseppe. *Arithmetices Principia*. Turin, 1889.

Plautus, Titus Maccius. *Trinummus*. In T. Macci Plauti, *Comediae*, vol. 2, ed. W. M. Lindsay. Oxford, 1903; 1963.

Poggendorff, Johann Christoff, ed. *Annalen der Physik und Chemie*. Leipzig, 1824.

―――. *Biographisch-Literarisches Handwörterbuch zur Geschichte der Exakten Naturwissenschaften*. Vol. 1. Leipzig, 1863; Amsterdam, 1965. Continuing series.

Poincaré, Jules Henri. *Science et l'Hypothése*. Paris, 1902. Translated by George Bruce Halsted, under the title *Science and Hypothesis*. Intro. by Josiah Royce. New York, 1905.

Poncelet, Jean Victor. *Traité des propriétés projectives des figures*. 2 vols. 2d ed., Paris, 1865–66.

Porphyry. *Porphyrii Isagoge et in Aristotelis Categorias Commentarium*. Edited by A. Busse. Berlin, 1887.

―――. *Isagoge*. Translated by Edward W. Warren. Toronto, Ontario, 1975.

Prihonsky, Frantisek. *Neuer Anti-Kant, oder Prüfung der Kritik der reinen Vernunft*. Bautzen, 1850.

Proclus. *In primum Euclidis elementorum librum commentarii*. Edited by G. Friedlein. Leipzig, 1873; Hildesheim, 1967. Translated by Glenn T. Morrow, under the title *A Commentary on the First Book of Euclid's Elements*. Uses the pagination of the Friedlein edition in the margins for reference. Princeton, 1970.

―――. *The Philosophical and Mathematical Commentaries*. *2 vols*. Translated by Thomas Taylor. London, 1792.

Putnam, Hilary. See Paul Benacerraf.

Rabel, Gabriele. See Kant.

Reichardt, Willi. "Kants Lehre von den synthetischen Urteilen apriori und ihre Bedutung für die Mathematik." *Philosophische Studien* 4 (1880): 595–639.

Reimarus, Johann Albert Heinrich. *Ueber die gründe der menschlichen erkenntniss und natürlichen religion*. Hamburg, 1787.

Remes, Unto and Jaakko Hintikka. *The Method of Analysis: Its Geometrical Origin and Its General Significance*. Boston, 1974.

Rickert, Heinrich. *Der Gegenstand der Erkenntnis: Einführung in die Tranzendental-Philosophie*. 3d ed. Tübingen, 1915.

―――. *Kulturwissenschaft und Naturwissenschaft*. 3d ed. Tübingen, 1915.

―――. *Allgemeine Grundlegung der Philosophie*. Tübingen, 1921.

Riehl, Aloys (Alois). *Der philosophische Kritizismus und seine Bedeutung für die positive Wissenschaft.* 2 vols. in 3 pts. Leipzig, 1876–87; 3d ed., Leipzig. 1926.

Risse, Wilhelm. *Die Logik der Neuzeit.* Vol. 1 (1500–1640). Stuttgart, 1964.

Ritzel, Hermann. *Über analytische Urteile: eine Studie zur Phänomenologie des Begriffs.* Inaugural Dissertation. Halle, 1916.

Rome, Adolphe. "Procédés anciens de calcul des combinaisons." *Annales de la Société Scientifique de Bruxelles* 50 (1930) series A (Sciences Mathématiques).

Rost, Johann Leonhard. *Astronomisches Handbuch.* Nürnberg, 1726.

Royal Society of London, Académie des Sciences, Paris, and others. *Neue Anmerkungen über all Teile der Naturlehre, aus denen englischen Transactionen, denen Gedenkschriften der Akademie der Wissenschaften in Paris, und andern mehr zusammengezogen und gesamelt. Zweiter Theil. Aus dem französischen übersetzt.* Kopenhagen and Leipzig, 1754.

Royce, Josiah. "Kant's Doctrine of the Basis of Mathematics." *Journal of Philosophy, Psychology and Scientific Methods.* 2 (April 1905): 197–207.

Rudnick, Hans H. "Translation and Kant's *Anschauung, Verstand,* and *Vernunft.*" In *Interpreting Kant,* ed. Moltke S. Gram. Iowa City, 1982.

Rudolph, Daniel Gottlob. *Anfangsgründe der Arithmetik, Geometrie und Trigonometrie.* Leipzig, 1757.

Russell, Bertrand, and Alfred North Whitehead. *Principia Mathematica.* 2d ed., Cambridge, 1925.

Saccheri, Giovanni Girolamo. *Euclides ab omni naevo vindicatus; sive conatus geometricus quo stabiliuntur geometria principiae.* Mediolani, 1733.

Sarganeck, Georg. *Die Geometrie in Tabellen.* Halle, 1759.

Savile, Henry. *Praelectiones tresdecim in principium Elementorum Euclidis.* Oxford, 1621.

Schepler, Herman C. "The Chronology of Pi." *Mathematics Magazine* (1950): 165–70, 216–28, 279–83.

Schleiermacher, Friedrich Ernst Daniel. *Dialektik.* Edited by L. Jonas. 1839; new critical ed., edited by I. Halpern, Berlin, 1903.

Schmalenbach, Herman. *Leibniz.* München, 1921.

Schmeisser, Friedrich. *Anleitung zum Selbstfinden der reinen Mathesis.* Berlin, 1817.

———. *Lehrbuch der reinen Mathesis.* Berlin, 1817.

Schmid, Anne-Françoise. *Une Philosophie de savant: Henri Poincaré et la logique mathématique.* Paris, 1978.

Schmid, Karl Christian Erhard. *Wörterbuch zum leichteren Gebrauch der Kantischen Schriften.* Jena, 1786; 4th ed., 1798.

Schönberger, Andreas. *Grundlinien zeiner Grössen-Wissenschaft in ihrer Natur dargestellt.* Vienna, 1801.

Schott, Gaspar. *Organum mathematicum libris IX. explicatum.* Herbipoli, 1668.

Schröder, Ernst. *Lehrbuch der Arithmetik und Algebra.* Vol. 1. Leipzig, 1873.

Schubert, Friedrich Wilhelm. *Immanuel Kants Biographie.* In *Kants Werke,* ed. K. Rosenkranz and F. W. Schubert, vol. 11, pt. 2. Leipzig, 1842.

Schubert, Hermann. *Elementare Arithmetik und Algebra.* Leipzig, 1899.

Schultz, Johann. *Entdeckte Theorie der Parallelen nebst einer Untersuchung über den Ursprung ihrer bisherigen Schwierigkeit.* Königsberg, 1784.

———. *Erläuterungen über des Herrn Professor Kant Critik der reinen Vernunft.* Königsberg, 1785; 2d ed., 1791; *im Gewande der Gegenwart* (in modern terms), new ed. by R. C. Hafferberg, Jena and Leipzig, 1898. (The same title is to be found under the author's name spelled "Johann Schulze".)

———. *De geometria acustica nec non de ratione seu basi calculi differentialis.* Inaugural Dissertation. Königsberg, 1787.

———. *Versuch einer genauen Theorie des Unendlichen.* Königsberg and Leipzig, 1788.

———. *Prüfung der Kantischen Critik der reinen Vernunft.* 2 pts. Königsberg, 1789–92.

———. *Anfangsgründe der reinen Mathesis.* Königsberg, 1790.

———. *Kurzer Lehrbegriff der Mathematik.* Königsberg, 1797.

———. *Sehr leichte und kurze Entwickelung einer der wichtigsten mathematischen Theorien.* Königsberg, 1803.

Schulze, Carl Ludwig. *Dissertatio Inauguralis exhibens nonulla ad doctrinam judiciis analyticis atque syntheticis spetantia.* Francofurti ad Viadrum, 1793.

Schulze, Gottlob Ernst. *Kritik der theoretischen Philosophie.* 2 vols. Hamburg, 1801.

Schulze, Johann Karl. *Taschenbuch der Messkunst.* Berlin, 1782.

Schumacher, Heinrich Christian. See Gauss.

Schütz, Christian Gottfried. *Programma de syntheticis mathematicorum pronuntiationibus.* 1785. Reprinted in *Opuscula philologica et philosophica.* Halle, 1830.

Schwab, Johann Christoph. *Ausführliche Erläuterung der von der Königlichen Akademie der Wissenschaften Berlin für das Jahr 1791 vorgelegten Frage:* "Welches sind die wirklichen Fortschritte, die die Metaphysik seit Leibnitzens und Wolffens Zeiten in Deutschland gemacht hat?" Berlin. 1796; Darmstadt, 1971.

———. *Commentatio in primum elementorum Euclidis librum, qua veritatem geometriae principiis ontologicis niti evincitur, omnesque propositiones, axiomatum geometri corum loco habitae, demonstrantur.* Stuttgart, 1814.

Segner, Johann Andreas von. *Deutliche und vollständige Vorlesungen über die Rechenkunst und Geometrie.* Lemgo, 1747; 2d ed., 1767.

———. *Elementa Arithmeticae, Geometriae et Calculi geometrici.* Halle, 1756; 1767.

———. *Anfangsgründe der Arithmetik, Geometrie und der geometrischen Berechnung* (from the Latin by J. W. Segner). Halle, 1764; 2d ed., 1767.

Semler, Christian Gottlieb. *Astrognosia nova.* Halle, 1742.

Smith, David Eugene. *History of Mathematics.* 2 vols. Boston, 1923.

———. "The History and Transcendence of Pi." In *Monographs in Modern Mathematics,* ed. J. W. A. Young. New York, 1911.

———, ed. *A Source Book in Mathematics.* 2 vols. New York, 1929; 1959.

Stahl, Konrad Dietrich Martin. *Zahlenarithmetik und Buchstabenrechnung.* Jena, 1797.

Steiner, Jakob. *Systematische Entwickelung der Abhängigkeit geometrischer Gestalten*. Berlin, 1832.

―――. *Die geometrischen Konstruktionen, ausgeführt mittels der geraden Linie und eines festen Kreises*. Berlin, 1833.

Sternfield, Robert. *Frege's Logical Theory*. Carbondale, Ill., 1966.

Stiefel, Michael. *Arithmetica integra*. Norimbergae, 1544.

Struensee, Carl August. *Anfangsgründe der Artillerie*. Leipzig and Liegnitz, 1760.

Struik, D. J., ed. *A Source Book in Mathematics, 1200–1800*. Cambridge, Mass., 1969.

Sturm, Leonhard Christoph. *Le veritable Vauban se montrant au lieu du faux Vauban*. A la Haye, 1710.

―――. *Architectura Militaris hypothetic-eclectica*. Wien and Nürnberg, 1755.

Succov, Laurenz Johann Daniel. *Erste Gründe der Bürgerlichen Baukunst in einem Zusammenhange und auf Verlangen entworfen*. Jena, 1751.

Sully, Henry. *Unterricht von der Eintheilung grosser und kleiner Uhren*. 2d ed., Lemgo, 1754.

Thibaut, Bernhard Friedrich. *Grundriss der reinen Mathematik*. Göttingen, 1801.

Thiele, Günther. *Wie sind die synthetischen Urteile der Mathematik apriori möglich?* Dissertation. Halle, 1869.

Tiedemann, Dietrich. "Über die Natur der Metaphysik, zur Prüfung von Herrn Professor Kants Grundsätzen." 3 pts. *Hessische Beiträge zur Gelehrsamkeit und Kunst*. Vol. 1. Frankfurt, 1785.

Tóth, Imre. "Das Parallelproblem im corpus Aristotelicum." *Archive for the History of the Exact Sciences* 3 (1966–67): 249–422.

―――. "Non-Euclidean Geometry before Euclid." *Scientific American* 221 (1969): 87–92.

Vaihinger, Hans. *Kommentar zu Kants Kritik der reinen Vernunft*. 2 pts. Stuttgart, Berlin and Leipzig, 1881–92; 2d ed., edited by R. Schmidt, 1922.

Vivanti, Giulio. "Infinitesimalrechnung." In *Vorlesungen über Geschichte der Mathematik*, ed. M. Cantor. 2d ed. Vol. 4, pt. 26. Leipzig, 1908; Stuttgart, 1965.

Vlaque, Adrian. *Trigonometria artificialis*. Grovdae, 1633.

Vloemans, Antoon. *Anschauung und Verstand in der Entwicklung von Kants Theorie der Geometrie unter Berücksichtigung von Descartes, Leibniz und Gauss*. Dissertation. Göttingen, 1921. Extract in *Jahrbuch der Philosophie*, Hälfte 1, 1, pp. 83–88, Fakultät Göttingen, 1921.

von Neumann, Johann (John). "The Formalist Foundations of Mathematics." Translated by Erna Putnam and Gerald J. Massey. In *Philosophy of Mathematics*, ed. P. Benacerraf and H. Putnam. Oxford, 1964; Englewood Cliffs, N. J., 1964. Originally in *Erkenntnis*, 2 (1931).

Wallerius, Johann Gottschalk. *Mineralogie, Oder Mineralreich, von Ihm ein-*

geteilt und beschrieben. 2d enlarged ed., Berlin, 1763.
Warda, Arthur. *Immanuel Kants Bücher, Bibliographien und Studien* 3, ed. M. Breslauer. Berlin, 1922.
Weidler, Joann Friederic. *Institutiones mathematicae.* Vittembergae, 1718.
———. *Tractatus de machinis hydraulicis.* Vittembergae, 1728.
Weischedel, Wilhelm. See Norbert Hinske.
Werckmeister, Andreas. *Erweiterte und verbesserte Orgel-Probe.* Quedlinburg and Aschersleben, 1716.
Whitehead, Alfred North, and Bertrand Russell. *Principia Mathematica.* 2d ed., Cambridge, 1925.
Wolff, Christian. *Anfangsgründe aller Mathematischen Wissenschaften.* 4 pts. 1710; 5th ed., Frankfurt, Leipzig, and Halle, 1750.
———. *Elementa matheseos universae.* 2 vols. Halle, 1713–15.
———. *Philosophia rationalis, sive Logica.* Frankfurt and Leipzig, 1728; 3d. ed., Frankfurt and Leipzig, 1740.
———. *Philosophia prima, siva Ontologia.* Frankfurt and Leipzig, 1729, and 1730; 2d ed., Frankfurt and Leipzig, 1736; Hildesheim and Darmstadt, 1962.
———. *Auszug aus den Anfangsgründen der Mathematischen Wissenschaften.* Frankfurt and Leipzig, 1749.
Wubnig, Judy. "7 + 5 = 12: Kant and Plato." *5. Internationaler Kant-Kongress: Akten, 1,* 71–79. 1981.
Wundt, Wilhelm. *Logik.* 2 vols. Stuttgart, 1880–83; 4th and 5th eds. in 3 vols., Stuttgart, 1919–24.

Zimmermann, Christian Gottlieb. *Entwicklung analytischer Grundsätze für den ersten Unterricht in der Mathematik.* Berlin, 1805.
———. *Anfangsgründe der Geometrie.* Berlin, 1814.
Zimmermann, Robert. "Über Kants mathematisches Vorurtheil und dessen Folgen." *Sitzungsbericht der phil.-hist. Classe der Kaiserlichen Akademie der Wissenschaften* 67 (1871): 7–48.

Name Index

Adam, Charles Ernest (1857–1940), 149, 153
Adickes, Erich (1866–1928), xix, 28, 141, 142
Aesop (ca. 619–546 B.C.), 67, 155
d'Alembert, Jean le Rond (1717–1783), xxvi
Allison, Henry E., 148
Apollonius of Perga (fl. 225 B.C.), 9, 17, 147
Archimedes (ca. 287–212 B.C.), 23, 43, 151
Aristotle (384–322 B.C.), vii, 6, 12, 145, 162
Arnoldt, Emil (1828–1905), xx, xxii, 142
Ashworth, E.J., xii
Aster, Ernst von (1880–1948), 112
Auwers, Karl Friedrich von (1864–1939), xi

Bacon, Francis (1561–1626), 155
Baltzer, Richard (1818–1887), 34
Barrow, Isaac (1642–1727), 13, 146
Baudouin de Guemaduc, Armand Henri (1737–?), xxvi
Beck, Hans (1876–1942), xviii, 142
Beck, Jacob Sigismund (1761–1840), 14, 25, 49, 69, 70, 75, 148, 152, 156
Becker, Oskar (1889–1964), xvii, 159
Beckmann, Petr, 153
Behmann, Heinrich (1891–1970), 114-15, 163
Belidor, Bernard Forest de (1698–1761), xxv
Bendavid, Lazarus (1762–1832), 102, 108, 161
Bennett, Jonathan, 146
Benvenuti, Carol (1716–1789), xxvi
Bergmann, Julius (1840–1904), 119, 163
Bering, Johann (1748–1825), 73, 156
Bernoulli, Daniel (1700–1782), xxvi
Bernoulli, Jacob (Jacques) (1654–1705), xxiii, 59, 154
Bessell, Friedrich Wilhelm (1784–1846), 8, 146

Beth, Evert W. (1908–1964), 157
Binkley, David, xii
Boerhaave, Hermann (1668–1738), 67, 155, 156
Boethius, Anicius Manlius Severinus (ca. 480–524), 155
Bolyai, Wolfgang (Farkas) (1775–1856), 37, 39, 150
Bolzano, Bernhard (1781–1848), 87, 110, 116–18, 121, 125, 159, 163
Borelli, J. A. (Giovanni Alfonso) (1608–1679), 12
Braunmühl, Anton Edler von (1853–1908), 58, 154
Brentano, Franz (1838–1917), 119, 125, 163
Bretschneider, Carl Anton (1808–1878), 34
Brouwer, Luitzen Egbertus Jan (1881–1966), 3, 114, 145
Buchenau, Artur (1879–1946), 146
Buck, Friedrich Johann (18th cent.), 142
Büttner, Christoph Andreas (1708–1774), xxiv

Cajori, Florian (1859–1930), 153
Cantor, Moritz (1829–1920), 45, 151, 154
Carnap, Rudolf (1891–1970), xviii, 94, 142
Cassirer, Ernst (1874–1945), 122, 146
Cicero, Marcus Tullius (106–43 B.C.), 155
Clavius, Christopher (1537–1612), 4, 7, 12, 13, 20, 32, 35, 149
Cohen, Hermann (1842–1918), 122–23, 125, 164
Couturat, Louis (1868–1914), xviii, xix, 22, 23, 35, 45, 61, 101, 114, 121, 142, 148, 155
Crelle, August Leopold (1780–1855), 47
Crusius, Christian August (1715–1775), 23-25, 27, 28, 30, 72, 148

Subject Index